YOUR
CAT

NATURALLY

YOUR
CAT
NATURALLY
A Guide to Behavior and Health Care

GRACE McHATTIE

With Natural Remedies by
Tim Couzens B. Vet Med. M.R.C.V.S. Vet M F Hom.

Special photography by Jane Burton

Carroll & Graf Publishers, Inc.
New York

First published in the United States in 1992 by
Carroll & Graf Publishers, Inc.

Carroll & Graf Publishers, Inc.
260 Fifth Avenue
New York, NY 10001

Library of Congress Cataloging-in-Publication Data

McHattie, Grace
    Your cat naturally / Grace McHattie.
        p. cm.
    "An Eddison-Sadd edition."
    ISBN 0-88184-804-2 : $18.95
    1. Cats.   2. Cats--Behavior.   I. Title.
    SF447.M346 1992
    636.8--dc20   92-9288
                    CIP

AN EDDISON · SADD EDITION
Edited, designed and produced by
Eddison Sadd Editions Limited
St Chad's Court
146B King's Cross Road
London WC1X 9DH

Phototypeset in Bauer Bodoni by SX Composing Ltd, Rayleigh, England
Origination by Columbia Offset, Singapore
Printed and bound by Graficromo, S.A., Cordoba, Spain

Page 1. *Cats use their whiskers, or* vibrissae, *to touch
things. Whiskers also reveal the cat's mood. Here they
are pushed forward, showing anticipation. Drawing
the whiskers back shows aggression.*

Page 2. *Cats love sitting in the sun. It is an excellent
source of vitamin D, which they lick from their coats.*

Contents

Introduction

Is there such a thing nowadays as a natural cat? And if there is, is that what we want our cats to be? The answer must be yes if we wish our cats to be healthy, happy and well-adjusted to life in the 20th century. Lifestyles have changed dramatically for human beings during the last hundred years and have changed no less dramatically for our closest animal companions. Cats are no longer mere mouse-catchers but part of the family, yet often we are no closer to understanding them than we were several generations ago.

There is no lack of evidence that close contact with a cat can keep *us* healthy. We are all familiar with recent statistics which show that cuddling a cat will help us live longer, remain healthier, lower our blood pressure and dramatically reduce stress. Now new studies show that close contact with a pet even helps hospitalized patients to wean themselves off sleeping pills.

But are we as good for the cat as the cat is for us? The answer is that we can be if we try to understand and cater for its needs. That is what this book tries to do. It seeks to help our understanding by exploring feline body language and natural behaviour and by explaining how we can keep our cats healthy by offering them good nutrition, exercise and preventative health care.

As more people are attracted to alternative therapies such as homoeopathy, herbalism, and Bach flower remedies for their own ailments, so too are many cat owners turning to natural therapies for their pets. I am delighted to introduce the consultant to this book, Tim Couzens B. Vet Med. M.R.C.V.S. Vet MF Hom., as a leading authority in this field. As a Member of the Royal College of Veterinary Surgeons and also a fully-qualified Veterinary Member of the Faculty of Homoeopathy, he is one of a growing number of veterinary surgeons who embrace conventional as well as alternative medicine. Should you wish to find a qualified vet who is also conversant with natural therapies, you can do so by contacting the British Association of Homoeopathic Veterinary Surgeons or the International Association for Veterinary Homoeopathy (IAVH) at the address on page 117. Please do not bypass qualified veterinary help in favour of self-diagnosis.

I hope this book will give you a working knowledge of feline psychology and help you to recognize the natural behaviour of a healthy cat. Any deviation from the norm, however minor, may be an indication that the cat needs veterinary attention.

My best regards to you and to the natural cat in your life.

Grace McHattie

Left. *Cats will find a comfortable, sunny spot for a rest no matter how inconvenient it is for humans! The only time felines will sleep on their backs is when they are on familiar territory, feeling safe and relaxed.*

INTRODUCTION TO
Natural Health Care

Have you ever stopped to think about why your cat should become ill? We all think of illness as a nuisance, to be sorted out as quickly as possible, with the minimum of inconvenience. Perhaps we should be looking deeper and examining the signs more closely. Each cat is, after all, an individual, and each illness and its symptoms particular to that animal. Perhaps the body is trying to tell us something. Try thinking of illness as an indicator of disharmony within the body. It would make more sense to treat the problem in a natural way, using medicines that work in harmony with the body and not in opposition to it. Whilst modern medicines undoubtedly have their place, they can suppress symptoms and may cause side effects that are not altogether obvious at first. Above all they do not help restore the balance in a natural way. Fortunately, nature has provided us with many remedies that we can use to help.

Remedies in the Book

Throughout this book there are advice panels on the use of natural remedies. These are

broadly divided into homoeopathic remedies and herbal remedies, although some mention is also made of the Bach flower remedies and the use of essential oils. The lists are not by any means complete as there is not enough space in this book to cover every possible remedy that could be used. The remedies included are those most commonly found to be effective in treating the relevant conditions. Remember that true holistic medicine involves not only using natural remedies to treat illness but also feeding your cat a correct diet (see Chapter One) and taking a close look at its life-style.

Diagnosis of a problem is equally important as the treatment and for this reason it is always necessary to maintain regular contact with your vet. There is very little point in putting your cat's life in jeopardy by using an inappropriate treatment or incorrectly diagnosing a problem.

Before embarking on any of the treatments I would urge you to read the next section, which highlights the differences between homoeopathic and herbal remedies and helps explain the basic principles behind them. Do not wait until you are faced with an ill cat before you realize that some herbal remedies

---

Left. Half-closed eyes show drowsy contentment. Many cats are happiest when dozing in the sun.

need prior preparation or that your health-food shop does not stock the homoeopathic remedy that you need urgently.

The Principles of Homoeopathy

The concept of homoeopathy can be difficult to understand. It is often confused with herbalism and although some plant remedies are common to both, their underlying principles are very different.

Homoeopathy was known to the Greeks at the time of Hippocrates. Like so many words in the English language it is derived from the Greek 'homoios' meaning similar and 'pathos' meaning suffering. Together, these two words express the basic principle of homoeopathic medicine.

We owe much of modern homoeopathy to the work of the German physician Samuel Hahnemann. Hahnemann lived in Leipzig, in what was East Germany, enjoying an international reputation as doctor, scholar and chemist. Towards the end of the 18th century he began work translating old medical texts into German. He found himself in disagreement with the action of Cinchona bark or quinine, when used in the treatment of malaria. To test its effect he took the drug himself, to find that it produced symptoms resembling malaria.

It seemed possible that drugs that produced certain symptoms in healthy people could be used to treat patients exhibiting similar signs. It was from this that he formulated his law of similars, adopting the phrase, *simila similibus curentur*, let like be cured by like. As a scientist he tested his theories extensively, not only taking remedies himself but also testing them on his friends and followers. He carefully noted the results produced by each medicine, including physical, mental and emotional symptoms. These he compiled into a *materia medica*, listing what symptoms each substance could cause as an indicator of what it was capable of curing. Hahnemann's *materia medica* listed almost 70 remedies. Today we have the drug pictures of more than 2000 homoeopathic remedies from which to work .

Guided by his law of healing, Hahnemann initially used large doses of his remedies to treat patients. This often caused initial aggravation of the symptoms, even though the ultimate results were good. To avoid this he diluted them in an orderly fashion, finding that not only were any side-effects removed, but that their medicinal power was increased. The more dilute he made his remedies the more potent they became. In this way potentially poisonous substances (in their natural state), such as rattlesnake venom or deadly nightshade can be used to treat illness safely. In much the same way seemingly inert substances such as silica or common salt can be put to therapeutic use.

Homoeopathic Doses

Homoeopathic remedies have to be diluted in a specific way for them to become effective. Almost any known substance can be prepared homoeopathically. The commonest method involves making an alcoholic extract of the original material, called the 'mother tincture' and represented by the symbol $\emptyset$. From this stage potentization of the remedy is carried out. This involves making a number of serial dilutions with vigorous shaking (known as succussion) at each stage. Paradoxically, the more dilute the mother tincture becomes, the more potent the remedy becomes homoeopathically. Homoeopathic remedies are commonly available in three potency ranges: x (decimal) where the dilutions are carried out in steps of 1 in 10, c (centesimal) produced by dilutions of 1 in 100 and M, where 1M equals 1000c. Arnica 6c, for

Cats are not solitary creatures and friendship is important to them. They like to have a companion and do not mind whether this is a fellow feline or a human.

example, is a 1/1,000,000,000,000 dilution of Arnica mother tincture. A homoeopathic pharmacist would produce this by initially taking 1 drop of Arnica ∅ and adding this to 99 drops of alcohol/water and shaking the mixture vigorously. He would then take 1 drop of the resulting solution (the 1c dilution), adding it to 99 drops of alcohol/water and shaking again. This would result in Arnica 2c. The process would be repeated a further 4 times to give Arnica 6c. A few drops of this would then be added to some plain lactose tablets, which would then become potentized. Potencies in the centesimal range are usually written without the suffix 'c', so Arnica 6 is the same as Arnica 6c.

The Value of Homoeopathy

Although scientifically speaking, when a remedy is diluted past 12c there should be no molecules of the original substance present, it continues to be effective. It is the process of shaking (succussion), carried out at each stage, that seems to release the healing energy inherent in each remedy, whilst dilution removes any toxic side-effects. It is thought that the remedies probably stimulate the body's own healing powers in some way, possibly by acting through the immune

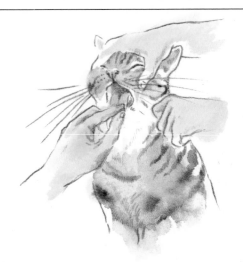

Tip your cat's head back and drop the tablet as far down the throat as possible. Homoeopathic remedies in tablet form should not be touched by hand but tipped from the container top.

Gently touch or blow upon the tip of the nose to make your cat swallow. If it does not swallow pills readily, hold the mouth closed without blocking the nose and stroke it under the chin.

system. Since we cannot measure the energy, as the remedies are so diluted and since the process appears unscientific, homoeopathy has met with considerable scepticism. Yet it is a remarkable and highly effective system of medicine if its principles are correctly applied. Another benefit of homoeopathic treatment is that, unlike conventional medicine, there are no side-effects, although symptoms occasionally appear to get worse before improving. Homoeopathic remedies can be used safely to treat the youngest kitten to the oldest cat, even those with kidney and liver problems. If the wrong remedy is given nothing happens.

Choosing a Remedy

When treating your cat homoeopathically for any problem, always bear in mind that the choice of remedy is crucial. The symptoms that are associated with the remedy must closely match the symptoms shown by the patient, otherwise it will fail to achieve a cure. External factors (known as modali-

ties) such as hot or cold, damp or dry or time of day and their effect on the patient or the patient's symptoms are also significant pointers in helping to identify the correct remedy. The choice of potency is also important. Lower potencies such as 3x have less energy, so the match does not have to be so exact. However, a lower potency means more frequent doses and weaker healing power. Whilst higher potencies may only need a few doses to achieve a cure, the match needs to be increasingly accurate the higher the potency used. The potencies given in the panels and the A-Z are only suggestions based on experience. Some cats may respond to one potency better than another, but in an emergency situation it is better to give any potency to your cat rather than none at all.

Administering Remedies

Remedies are commonly supplied as tablets, but are also available as granules, powders and liquids. As 'energized' products they need special care and handling. They are deactivated by strong smells, especially camphor, eucalyptus and the like. Heat, sunlight and electromagnetic radiation also have the same effect. Once depotentized they become ineffective. Try to avoid handling homoeopathic tablets as this can also affect them — the best way of dosing a cat is to tip a tablet on to the lid of the bottle, then straight into the cat's mouth. Where this is not possible hide the tablet in some food, although this can reduce the effectiveness. An alternative is to use powders tipped straight on to the cat's tongue where they should stick, but these are more expensive and it is usually cheaper to crush up a tablet between two sheets of clean white paper. Liquid potencies are not recommended as the alcoholic content tends to cause salivation.

Common remedies such as Arnica are available from most health-food shops and some chemists, usually as 6 or 30 potency. Other remedies or potencies will need to be obtained from a homoeopathic pharmacy (see Suppliers, page 117 for details).

The Origins of Herbal Medicine

Herbalism, sometimes called phytotherapy, is probably the oldest system of medicine in existence. Although known to Hippocrates, records dating back even earlier, to 2500 BC, mention the use of medicinal plants. It became universal to all peoples and races — plants are an integral part of our planet and of our lives.

By the time of Gerard, the famous herbalist of the 17th century, much had been written about the use of plants in treating illness. Gerard's *Herbal*, printed in 1636, gave the medicinal uses of some 3800 plants. As with homoeopathy, orthodox treatment became popular as the older forms of medicine were increasingly ridiculed. Herbalism, both human and veterinary, became associated with the bad practice of using herbs in large toxic doses (which caused side-effects) and with administering poisons such as arsenic and mercury. Its decline was accelerated by its link with astrology. This was unfortunate as the fundamental basis of herbalism has always been sound. Now we can reflect on the value of herbal treatments and the role of plants in our own and our pets' well-being. We have only to look at cultures such as the Chinese and Native American, where herbal medicine continues to flourish today.

The use of herbs to treat animals would seem logical. Cats living in the wild would instinctively go and seek plants to heal themselves when ill. They even have preferences for herbs such as couchgrass and catmint.

The Value of Herbal Medicine

Every medicinal plant is a mixture of components. Modern medicine has singled out some of these and put them to therapeutic use or made synthetic derivatives from them — Atropine from deadly nightshade, Aspirin from willow bark and Digitalis from the foxglove to name a few. Yet the isolated compounds are likely to be less effective. Their synthetic counterparts are costly to develop and often associated with unwanted side-effects. Herbal medicines rarely cause side-effects as all the components of an individual remedy are balanced as nature intended. They do not accumulate in the body and do not produce withdrawal symptoms. Many of the remedies also have more than one action. Using the whole plant results in an enhanced therapeutic effect as the various constituents may either potentiate each other in their actions or compensate for any untoward effects. Herbal remedies work in harmony with the body, not against it, gently and slowly assisting it to overcome the problem in a non-aggressive way.

Commercial Herbal Remedies

Remedies are available from several sources. Some companies produce specific herbal products for treating domestic pets. They are obtainable from pet shops and include a dose guide on the side of the packaging. If you look closely you will find that some products contain more than one herb. For example, you may see several herbs that all act on the urinary system combined together to help treat a wider variety of problems.

Home-made Remedies

One alternative to using the prepared veterinary herbal remedies is to make tinctures from dried herbs. You will have to do this for those remedies that you cannot obtain otherwise. Most tinctures for human use are

Trees are, in many ways, a natural habitat for your cat. They provide places to hide, shade from the sun and trunks to scratch. The bark, leaves and roots of some trees can provide medicine.

based on alcohol, which cats tend to find extremely unpalatable. It is far better to use glycerine, which may not be such a good solvent but is less likely to cause problems with dosing. To make a standard tincture add 25 ml glycerine to 25 ml water. Add 10 grams of dried, ground herb and pour into a well-stoppered bottle. Leave in a warm place, shaking well every day for 2 weeks, after which time you should decant off the liquid. Pour the remaining residue into a muslin cloth and wring out to recover all the liquid. Such a preparation will keep for several months.

Dried herbs can be also used to make infusions. This is a bit like making tea, you can even use a teapot if you wish. To make a standard infusion add half a cup of boiling water to half a teaspoonful of dried herb and leave to steep for about 15 minutes. Strain and then leave to cool. Where a herb is very woody it is better to make a decoction rather than an infusion. This will ensure that all the constituents go into solution. Use the same quantities as above, but add a little extra water to allow for evaporation. Boil and simmer for 15 minutes in a non-aluminium saucepan and strain while hot.

A Guide to Dosage
(for an adult cat of average weight)

Tinctures Five drops tincture three times daily. This is roughly equivalent to 1 drop tincture per kg body-weight, three times daily, added either to food or given by mouth. You will find mention in the panels and the A-Z of herbs that combine together well.

Infusions and Decoctions Three teaspoonfuls (15 ml) three times daily, mixed in with food if possible. This is approximately 3 ml per kg body-weight, three times daily. Infusions and decoctions, unlike tinctures, do not keep well and should be made up daily as required.

Compresses Some herbal infusions can also be put to use externally in the form of a compress – to help wounds heal or reduce bruising, for example. To make a compress, soak some clean cotton wool in a warm infusion. Place over the affected area and hold in place for five minutes, longer if possible. Repeat several times a day where needed.

Poultices These are similar to compresses, but use either fresh or dried herbs. If you want to use fresh herbs bruise the leaves and place directly on the affected part. Make dried herbs into a paste using hot water and apply to the skin. Poultices are of most value in drawing pus out of abscesses.

Bach Flower Remedies

Dr Edward Bach pioneered the 38 flower remedies that bear his name in the 1930s. The remedies, all prepared from the flowers of wild plants, bushes or trees, have no side-effects. They are prescribed on the state of mind and not directly for any physical complaint. Continued worry or grief can affect general health and vitality, leading to illness. This is much the same for cats. Each flower remedy is supplied as a stock solution preserved in brandy. This needs to be diluted for general use. You can use more than one remedy at a time if you wish. Make a treatment bottle using 30 ml of still spring water and add two drops of each chosen remedy and a little cider vinegar to act as a preservative. The normal dose is two or three drops four times daily. You may need to treat for some time in certain cases.

Essential Oils

Essential oils are vital essences extracted from aromatic plants. The flowers, leaves, bark, berries or roots are used as the source depending on the individual oil. Their uses are various, including the treatment of minor ailments and in helping to repel fleas. They should never be given internally to cats. Always use pure essential oils, never their synthetic counterparts, which are not as effective. It is also wise not to buy any oil sold already in a base carrier oil as these have very short shelf-lives. Advice on using essential oils in treatment is given, where applicable, throughout the book.

Before You Start

Before you begin any sort of treatment, always try to make sure that you know what the problem is. If there is any doubt in your mind, seek professional help – it is far better to be safe than sorry. If the problem is getting worse or is not responding to your choice of remedy, it is wise to seek advice. Whether or not you decide to ask for help will depend also on how familiar you are with cat ailments and how much you have learnt from using natural remedies. Always follow your instinct – giving an appropriate remedy, even on the way to the vet, can improve the outcome of a serious case.

Consulting Your Vet

Some conditions can only be diagnosed accurately by your vet. For example, some specific diseases such as feline leukaemia, feline infectious anaemia and feline infectious peritonitis and other problems such as diabetes, liver disease or kidney failure can only be confirmed by a blood test. X-rays are the only way to check for a fracture or a bowel obstruction and an anaesthetic may be needed to check for a foreign body lodged in the throat. Other conditions call for more than natural remedies alone. A very dehydrated cat, for example, may need intravenous-fluid therapy; a cat with feline lower urinary tract disease (previously called FUS) will need a low-mineral diet to help prevent reoccurrence; a cat with kidney failure will need a diet with the correct protein composition; and particular problems, like a large wound, can only be dealt with surgically.

Choosing your Remedy

When treating your cat at home, you will have to decide whether to use a herbal or homoeopathic remedy. Although it is possible to use both alongside each other, it is better to select one or the other at the outset, only changing if there is no response to the initial therapy. To some extent your choice will be dictated by your personal preferences, the remedies you have at hand and by your experience in using them. Herbal remedies are sometimes slow to act but their action is usually predictable, whilst homoeopathic medicines rely to some extent on your ability to match the symptoms to the remedy. Homoeopathic remedies can be given safely to any animal of any age. The dose will be the same for a new-born kitten as for the oldest cat. In contrast, if you plan to use herbal remedies you will have to work out the dose first, based on body-weight (see p. 15). It is best to seek guidance when either treating very young kittens or pregnant queens (unless the herbs are specifically recommended for use during pregnancy).

Phasing Out Conventional Treatment

You will need to give some thought to any conventional treatments your cat has received. Herbal medicines are unaffected by

Understanding your cat's natural needs – sleeping and sunlight are important requirements – will ensure your cat stays healthy and contented.

modern drugs, in fact herbal remedies can be used alongside conventional medicines without compromising the action of either. But if you plan to do this it would be wise to tell your vet. You might both be pleasantly surprised when the condition improves sufficiently to phase out the conventional drugs. Homoeopathic remedies are more sensitive. Garlic is known to affect homoeopathic remedies, but some conventional veterinary medicines (particularly steroids) can have a more profound effect, even to the extent of counteracting them completely. Always consult your vet if you plan to withdraw any long-term conventional therapy. Fortunately homoeopathic remedies will work alongside many modern medicines, without losing any of their effect.

Length of Treatment

Finally you will have to decide how long to treat for. Specific time-scales are given for some treatments in this book, but most acute problems (such as a bout of diarrhoea or cystitis) should respond in 24 to 48 hours. Sprains, strains, bites and other injuries should resolve in 10 days or so. Chronic problems such as kidney disease or arthritis will either need repeated courses or continuous treatment.

Successfully treating a problem with natural medicines can be both rewarding and satisfying. This is especially so where conventional medicine cannot provide the answer or seems unable to supply a satisfactory solution. Given a little time, patience and experience you can achieve good results.

Tim Couzens
B. Vet Med. M.R.C.V.S. Vet MF Hom.

CHAPTER ONE

A Healthy Diet

Survival is a cat's most basic drive. Living in the wild and having to fend for itself, it would eat a wide variety of animals. Insects and small rodents would form the basis of the diet supplemented with rats, rabbits, birds, birds' eggs and the occasional fish – if the cat was skilful enough to catch them. Scraps scavenged from dustbins or begged from kindly cat-lovers would make up the rest. Cats are great opportunists. The idea that they are fussy eaters is a myth. I have heard of owners who firmly believe their cat will eat nothing but best steak dipped in yeast extract. These same cats, given a can of ordinary cat food (by someone other than their owners) will tuck in with enormous relish. Cats will, simply, hold out for the best deal going.

Another misconception is that it is dangerous for a cat to eat insects. Many owners are horrified when they see their pets scoffing house flies, which we all know are a source of many kinds of infection. However, a cat's gastric juices are more than a match for the humble fly. Only if your cat's diet consists largely of grasshoppers need you worry – they cause constipation. Other insects not only give a cat a satisfying hunting sequence, but they also provide a sufficient natural reward when captured.

Cats also have an unfounded reputation for killing birds. Only 6 per cent or less of a wild-living cat's diet consists of birds – they are much harder to catch than mice, functioning, as they do, in three dimensions. Cats will generally catch only *slow* birds – those that are old or sick or fledglings that have fallen out of the nest and may have a short life expectancy anyway.

When hungry a cat will eat virtually *all* of its prey. The waste is minimal. After catching a bird, only a few feathers, the beak and the legs are left. The cat first tears out the larger indigestible feathers, then starts at the head because this is the way the feathers (or fur in the case of a rodent) lie. A snake also eats its prey in this way, following the 'lie' of the fur to avoid choking. This is why a well-fed house cat will eat the brains of a prey animal and leave the body. It finds it is not hungry enough to eat the entire carcass. In the wild, it would gorge on as much as it could eat, in

Left. A domestic 'tiger' prowls in the undergrowth, stalking its prey. Today the target is just an insect but the hunt is as intense as that of any big cat in the jungle.

case it does not catch anything for several more days. It would leave any uneaten food covered by earth or leaves for when it was next hungry. When a house cat makes scratching motions around a rather unappetizing food, it is 'covering it up' and saving it for a rainy day. It is not showing contempt for the food as many owners believe.

Cats eat by shearing meat into pieces with their incisor teeth, using them like knives; then they swallow these pieces virtually whole. In the stomach, food is mixed with acids and digestive enzymes, passing on to the small intestine where digestive juices from the liver and pancreas break down and absorb the food. Water is then reabsorbed

Fish is considered a traditional food for cats and poaching is a common sight on quaysides. But if a cat's diet consists mainly of raw fish, it could end up with a vitamin deficiency.

from the large intestine and waste products are eliminated. This is a relatively speedy process as the cat's intestine is just over 2 m (7 ft) long.

Having eaten its fill, the wild-living cat would need a drink. Although the prey animal's body would be approximately 80 per cent moisture, extra water is essential. A cat will find a water-hole, preferably large, with free-flowing water and at some distance from the area in which it chooses to defecate.

What is it that attracts a cat to its food? In the case of the wild-living cat, it is movement. A moving creature triggers a hunting instinct that is too powerful for any cat to resist. In the case of the house cat, fed on a commercially prepared diet, this trigger is first of all smell and then temperature and taste.

Cats have an organ (the Jacobson's organ) that allows a cat to both smell the air and 'taste' it at the same time. The cat keeps its mouth slightly open in a grimace as it draws in air over the roof of its mouth. This 'smell and taste' faculty explains why cats will not eat when they have a blocked nose. They have to be persuaded with strong-smelling food such as pilchards or mackerel, or by adding beef stock cubes or yeast extract to their meals. Cat food manufacturers tread a delicate line between producing foods that are strong-smelling enough to be irresistible to cats but not so strong-smelling that they repulse cat owners.

Taste is equally important. Most manufacturers spend large sums of money researching palatability, and many owners make the mistake of equating tastiness with

nourishment. If a cat prefers a particular food, it does not necessarily mean it is the most nourishing. After all, most of us would rather eat a chocolate than a stick of raw car-

Above and below. *Bowls should be large, as cats prefer them to be wider than their whiskers. If not, they may trail the food on the floor. Be sure to measure portions carefully. It is tempting for an owner to fill a bowl, whatever size it is.*

rot until we develop a taste for healthy food. Read labels to choose the best food for your cat and ignore those pleading looks.

Temperature is also important. Cats do not appreciate food straight out of the refrigerator although some greedy individuals will gobble it down. Most cats prefer food to be blood temperature – the temperature of freshly-caught prey. Refrigerated food can be warmed by leaving it at room temperature for a few hours or by putting it for a few moments in a microwave oven.

Cats are relatively unconcerned about texture, eating crunchy, dry food or kibble just as readily as soft, canned food. Appearance of food is totally unimportant. Many cat-food manufacturers add colourings to their food to make it look better to *us*. Your cat does not care if its food is rich brown or bright cerulean blue – as long as it smells and tastes right it will eat avidly.

Feeding Patterns

A cat should be fed at least twice a day. Urine remains more consistently acidic if it eats several small meals rather than just one and acidic urine is less likely to lead to urological problems. Do be sure that if you feed your cat twice a day instead of once, that you give it half its daily quota at one meal and not twice as much as it needs. For the average, neutered, adult cat, two-thirds of a 400 g (14 oz) can of an 'average' cat food a day is about right. Cats require approximately 80 to 90 calories for each kg of their body weight per day or 35 to 40 calories per lb per day. Decrease quantities slightly if your cat is becoming paunchy or defecating more than twice a day, and increase quantities slightly if your cat is trying to eat the goldfish.

At various times of its life, your cat will have increased nutritional needs. While pregnant or lactating she should be allowed as much food as she wants; she will probably eat three to four times as much as normal. Kittens also eat more than an adult neutered cat as they need extra energy to grow. Feed them four or five small meals a day but cut down the quantities if they have diarrhoea or if their tummies start to bulge. I know one person who was overfeeding his kittens so much that they were using their litter tray three or four times an hour. This is a sure sign of overfeeding.

The biggest influence on the domesticated cat's diet is undoubtedly its owner. Because cats are thought to be 'fussy' eaters, it can become a challenge to find a food the cat enjoys. It is a point of honour with some people that they alone know what their cat will eat. A hungry cat will eat virtually anything edible – it is not stupid. If owners want to change the diet – to reduce the liver content and so avoid vitamin A overdoses, or to introduce a more balanced combination of nutrients, for example – they may find it impossible to persuade their cats to eat other foods. This is probably because they are open to feline blackmail.

Vets, and those who run boarding catteries, have no difficulty in persuading a faddy cat to change its diet. They present the new diet and if the cat refuses to eat it, it goes hungry, though it rarely stops eating for more than a day or two. Cats can survive for weeks without food, as long as they have water to drink, so a day or two without food during a diet change will do no harm at all. In fact, a short fast can prove positively beneficial for a healthy adult cat. It cleanses the body of toxins, leaving the cat with bright eyes, glossy fur and loads of energy. So try not to give up too soon. Do not, however, allow a cat in kitten, a lactating cat, an old, ill or very young cat to fast, as it can injure its health.

Drinks are vital to a cat, but milk is not

It may look as if this cat has designs on the goldfish but it is just having a drink. The garden pond is the nearest most cats will get to the natural water-hole they would prefer.

necessary. Many have an allergic reaction to the lactose in cows' milk, so if your cat has recurring diarrhoea cut out its milk and see if the problem ceases. No mammal on a well-balanced diet needs milk once it has been weaned. Some cat-food manufacturers are now producing low-lactose milk specifically for cats. Again, this is probably more for the benefit of owners who believe cats need and deserve milk, than it is for cats. But by all means give it to them if it makes you happy. Nursing cats may appreciate milk for the calcium it contains. Powdered milk specifically made for weaning kittens is the most suitable kind to give.

Water is the liquid that is really essential for cats. One cat survived being locked up in a room accidentally for six weeks because it was able to lick condensation from the surrounding walls.

Since a cat would drink from a large water-

hole in the wild, a little plastic bowl will not seem very appealing. In fact, most cats drink more water than their owners realize, even if it is from a muddy puddle. To encourage your cat to drink, give it a large receptacle, such as a fish tank or bowl. Many cats already drink from fish tanks, not, as sceptical owners believe, because they are sneaking up on Goldy, but simply because a large tank is more like a natural water-hole than anything else they are given to drink from.

Cats are sensitive to the chemicals in tap water. You could try bottling it without a lid for a day or two in the refrigerator before pouring it into a bowl to give time for the chemicals to disperse or use uncarbonated bottled spring water.

Alternatively, you could tempt your cat with a little honey added to the water, or with stock made from boiled meat scraps and bones, drained and cooled. Most cats will not want to eat or drink near their litter tray, so site this some distance away, preferably in another room.

Have you ever noticed your cat tap the surface of its drinking water before drinking? This is not a ploy to tramp wet footprints all

Encourage a cat to drink by giving it a large container. This cat is enjoying using the terracotta base from a large flower pot but you could also use a washing-up basin, baby bath, fish tank or bowl (without fish). Some owners never see their cat drinking because it prefers to quench its thirst from a puddle, water-butt or stream, rather than the little dish most are given. Cats can obtain up to 85 per cent of the moisture they need from fresh or canned food, but clean water should still be available.

over your clean floor. By making the surface of the water vibrate, a cat can gauge how far away it is. That way, it doesn't get its nose wet when drinking!

Food Choices

It has never been easier to feed a cat a scientifically balanced diet. Most owners nowadays give their cats commercially prepared foods that contain optimum quantities of all known required nutrients.

Protein This provides the building blocks — amino acids – which promote growth, help to regulate the body's natural functions and renew body tissues while also supplying energy. The cat can synthesize ten amino acids in its body but a further ten, called the 'essential' amino acids, must be obtained directly from the diet. Foods with the most essential amino acids are eggs, milk, meat and fish.

Carbohydrates These starches and sugars are the fuel that produces energy. They are not essential as dietary fat can perform this function for a cat. Added carbohydrate, found in some cheaper cat foods, may slow down digestion and the passage of food through the gut.

Fat A concentrated source of energy that also transports vitamins A, D, E and K around the body. It is also needed as a source of the essential fatty acids (EFAs) some of which occur only in animal fats or fish oils. Added fat makes the diet more palatable and attractive to a cat in the same way that spreading butter on bread makes bread more palatable to us. Cats will eat huge chunks of fat with great relish and, as they do not suffer from coronary artery disease, this should do them no harm.

Vitamins These are necessary to promote and regulate biological and chemical processes in the body. The fat-soluble vitamins A, D, E and K can be stored in the body so a daily intake is not essential. An overdose can cause toxicity. Brewer's Yeast tablets are a good source of the water-soluble B vitamins: thiamin, riboflavin, pantothenic acid, nicotinic acid, pyridoxine, biotin, folic acid, choline and vitamin B_{12}. These vitamins are excreted in the urine if taken to excess and it is unlikely that an overdose of the water-soluble vitamins could be given. Cats do not need a source of vitamin C as they can synthesize it from glucose, but it is sometimes prescribed in cases of Feline Urological Syndrome to acidify the urine.

Minerals These regulate the body. The mineral content of cat food is shown on the label as 'ash'. Ash is what remains when the food is burned and consists of the macrominerals: calcium, phosphorus, potassium, sodium and magnesium, and the microminerals: iron, zinc, copper, manganese, iodine and selenium.

Both vitamins and minerals are usually added to prepared pet foods in carefully measured quantities to replace those lost in the cooking process or to add to the nutritional value of inferior constituents.

There are three basic types of prepared cat foods – canned, dry and semi-moist. Owners in the UK prefer to feed their cats mainly on canned foods and buy very little semi-moist, while those from other countries choose the semi-moist and dry foods. Nutritional quality varies little between different types of foods, so owner preference and convenience can dictate the choice.

Canned foods are available in a wide variety of flavours and textures. They range from high-quality, super-premium or gour-

HEALTHY EATING

Food is often an emotive subject for cat owners. Cats are considered to be fussy eaters, but this is not always true. Often owners confuse feeding with love. If a cat refuses a meal, however, it does not mean that the cat does not love its owner, it may simply be that it is not hungry, or does not enjoy the food on offer. The weather may be too hot, the food too cold, or the cat may be feeling ill. Owners must learn not to take the rejection of food personally and to accept that it is normal for a cat's appetite to vary.

Following a few simple rules should ensure your cat enjoys its mealtimes. Choose a good-quality product and feed your cat twice a day. Put the food into your cat's bowl then ignore it and your cat. Many owners are unable to change their cat's food because they fall for the feline histrionics that ensue when the new food, although more nutritious, is less palatable. Read the labels. And having rationalized your choice, stick to it.

This cat is using all its body language to express interest, from its pricked-up ears to its curved tail. Its paws are reaching up, almost as if to take the bowl from its owner's hands, and it rises on its hind legs to get closer to the bowl.

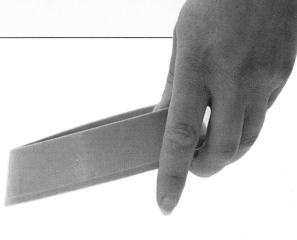

FOOD CHOICES

The most important thing to remember about food choices is that you have one but your cat does not. Although an outdoor cat may be able to make up any deficiencies by hunting, an indoor cat is totally dependent upon your choices. Be a good owner: read labels and decide what food is best for your cat, then buy it, whatever the price or personal inconvenience.

CANNED FOOD

- Very wide variety of flavour, quality and price for the owner to choose from.
- High moisture content: owners need not worry so much about the amount their cat drinks.
- Most cans are recyclable (read the labels).

- Content varies drastically, so always read can labels very carefully.
- Sizes are not always convenient for single-cat households.
- Heavy to carry: buy in bulk and have delivered if possible (it is cheaper too).

SEMI-MOIST FOOD

- As with cans, refrigeration is not necessary until the packet is opened.
- Inexpensive and often packed in convenient one-meal sizes.
- Said to produce a more concentrated form of faeces due to low moisture content.

- Choose non-plastic, recyclable packaging where available.
- Ensure cats drink more water when fed this type of food.
- A specialized product that has not yet taken off in some countries.

DRY FOOD

- Good exercise for teeth and gums – sprinkle some on top of other types of food.
- Does not 'go off' or require refrigeration.
- Usually an inexpensive food with a wide variety of flavours.

- It is essential that cats on a 'biscuit' or kibble diet drink extra water.
- Ensure that the brand you feed is nutritionally 'complete'.
- Can be smelly, especially if stored in bulk.

FRESH FOOD

- With fresh foods, you have more idea what you are feeding your cat than with some prepared foods.
- Owners enjoy a sense of achievement when preparing a delicious meal that their cat enjoys.

- More expensive to purchase than prepared foods.
- More time-consuming to prepare than other foods.
- For a balanced diet, feed meat, fish, vegetables, grains, dairy produce and pulses.

met foods, which are usually high in protein and rich in the meat or fish named on the label, to budget foods that can contain little or virtually nothing of the named meat. Some cheaper foods consist of offal, dried blood and fish trimmings combined with cereals, vegetables, pulses and oil seeds. The resulting gloop is flavoured, coloured and packaged and, if the chemists have done their job properly, can taste very good and be thoroughly enjoyed by the cat. Canned foods contain approximately 65 per cent to 85 per cent moisture and so are closest in moisture content to a cat's 'natural' prey diet.

The contents of semi-moist and dry foods are similar to that of canned. Semi-moist food is soft and easy to chew and older cats sometimes prefer it for that reason. It is usually packaged in plastic containers and contains from 25 per cent to 40 per cent moisture. Water or gravy can be added to it if a cat appears not to be drinking enough.

There are several misconceptions surrounding dry food. 'Biscuits' or kibble are complete foods and a cat can remain healthy on an exclusively dry-food diet. However, always check the label and ensure that what you are feeding is a 'complete' food; fish or meat must be added to some meal-based dry foods. Some owners mistakenly believe that dry foods are particularly high in protein. The label may state that a dry food contains, for example, 30 per cent protein but this is on a *dry-matter basis*. The following sum will reveal a true comparison of protein levels.

If a canned food has a protein level of 8 per cent and contains 80 per cent water the protein content on a dry-matter basis is found by subtracting the water content (80 per cent) from 100 giving 20. Divide the protein level (8 per cent) by 20 giving 0.4 and multiply this by 100 giving 40 per cent [8 divided by $(100 - 80) \times 100$].

So a protein level of 40 per cent in some dry cat foods will equate to a protein level of 8 per cent in some canned cat foods. Dry foods contain 5 to 10 per cent moisture so owners should check that a cat fed on dry food drinks plenty of water. If the cat refuses to drink or to eat canned foods, dry food can be moistened with water or gravy. A few dry biscuits can be sprinkled on top for a crunchy bite.

Today, very few owners feed their cats entirely on home-prepared food. It is much simpler to open a package or can than to cook a balanced meal. It is also cheaper. Over the past few years, the cost of meat and fish has increased several times more than pet food. Cats fed on home-prepared meals often become faddy, insisting on eating only one type of fish – usually the most expensive – or refusing to eat anything but liver.

One of the pitfalls of home-prepared food is the temptation to feed offal because it is cheap. Heart is not a good nutritional source and is high in magnesium; melts (spleen) can cause diarrhoea; and lights (lungs) have almost no nutritional value at all. Liver, the offal most commonly fed to cats is a frequent source of addiction, and its high vitamin-A content can lead to distorted bones, lameness, stiffness, gingivitis and tooth problems – a condition known as hypervitaminosis A. In some cases, bony outgrowths have to be removed surgically. Liver should never form more than 10 per cent of a cat's diet, although foods *flavoured* with liver may contain very little and could be fed more often. 50 g (2 oz) liver once a week will give a cat its requirement of vitamin A.

Fish (white and oily) is a good nutritional source although it is not really a 'natural' diet for a cat as few would bother – or be able – to catch it for themselves. It should be cooked, since raw fish contains an enzyme, thiaminase, which inactivates thiamin (vitamin

Many cats enjoy eggs but these should always be lightly cooked, preferably scrambled, as there might be a risk of salmonella poisoning from raw or undercooked eggs.

B_1). Eggs should be cooked for a similar reason; raw egg-white contains avidin which prevents another B vitamin, biotin, from becoming absorbed into the bloodstream.

Some home-cooked diets may not be ideally balanced. If, for example, a cat eats a great deal of white fish, it may need a feline vitamin and mineral supplement – obtainable from a vet. Those fed on a ready-prepared complete food, however, rarely need a supplement, except occasionally in the case of kittens and nursing female cats. Always

consult your vet before giving supplements.

Cats do not generally thrive on a vegetarian diet. They are carnivores, and need meat, fish or shellfish as a direct source of taurine. Taurine is an amino-acid-like substance whose importance has only been recognized relatively recently. Dogs can synthesize sufficient quantities for their needs but cats *must* obtain it directly from their diet. Without sufficient taurine, a cat will develop retinal lesions leading to irrevocable blindness, degeneration of the heart muscle and eventual death. It may take time, years perhaps, for symptoms to show, but they are inevitable if the diet lacks this essential nutrient.

A good, ready-prepared, complete cat food will meet a cat's every nutritional need. But it is important to read the labels carefully. Supplementary foods are incomplete whilst snacks are simply treats; neither provide all the necessary nutrients and should form only part of a cat's diet. The label also gives information about additives. These can include vitamins and minerals, colourings, flavourings, antioxidants (to prevent fat from becoming rancid), emulsifiers (to stop water and fat from separating) and antimicrobial agents (to slow down spoilage). Additives used in pet foods are the same as those used in human food, therefore pet-food manufacturers claim they are safe. However, people concerned about additives in their own food may wish to cut down those taken by their cat. There are several brands now on the market with fewer additives, and some with virtually none.

It is important for cats to become accustomed to a variety of foods from an early age because if a favourite brand or flavour becomes unavailable, a cat may refuse to eat. But be careful not to change brands too suddenly since this can cause diarrhoea.

If you choose your brands carefully when your cat is a kitten, you should have no difficulties later in life. Pick a good-value food with a protein level of at least 7½-8 per cent, preferably derived from meat, fish or their by-products, since animal protein is of a higher quality than vegetable. Pick a food with a low ash level too, for the average adult cat, since high mineral levels can cause urological problems. By careful comparison you will see that the most expensive is not always the best — the price may be high to pay for advertising.

If, for any reason, your cat is proving resistant to a change of diet, do not be discouraged by a refusal to eat for a day or two. Many authorities advocate a one-day fast each week for pets to allow the digestive system to cleanse itself of impurities (see p. 24). If you decide to follow this, choose the same day each week as cats are creatures of habit. Make sure there is plenty of water available, and that neighbours do not offer food. A film may form on your cat's teeth as toxins are eliminated from the system, so fast day is a good day to clean the teeth.

Feeding bowls need to be out of the main traffic areas, so that the cat can eat undisturbed and at some distance from the litter tray. Clean bowls with a solution of bleach (one without citrus or pine perfumes, which will repel your cat), or with a dilution of hydrogen peroxide, and rinse and dry well. If you have several cats, use sodium hypochlorite to prevent the spread of any infection.

Whether you choose a ceramic, stainless steel or plastic bowl, ensure that it is large enough. It needs to be wider than your cat's whiskers as cats dislike narrow bowls and may trail the food on to the floor to eat there. More sophisticated styles are now available. 'Hoppers', for example, hold dry food that pours into a bowl as the cat empties it. Others incorporate timers so that working owners can ensure their cat does not miss its lunch.

Trouble-free Toileting

Most cats do not need to be house-trained. When a kitten takes its first wobbly steps outside its nest at three to four weeks it will often climb straight into its litter tray. After eating some litter (like human babies, kittens put everything into their mouth) it will use the tray for its appointed purpose.

Kittens have small bladders, so will need to urinate (and defecate) several times a day. An adult cat will urinate, on average, twice a day, and will void from 22 ml to 30 ml (¾ to 1 fl oz) for each 1kg (2⅕ lbs) of body-weight each day. However, as urine is used as a territorial marker, some assertive cats may uri-

NATURAL FOOD SUPPLEMENTS

Read 'Before You Start', pp. 16-17

All the vitamins, minerals and trace elements that your cat needs can be provided by using some of the following natural supplements.

Seaweed (Kelp, *Fucus vesiculosus*)
Rich in vitamins, minerals and amino acids. Due to its iodine content it stimulates the thyroid gland, encouraging coat growth and good pigmentation. A valuable addition for overweight cats, helping to use up calories by stimulating the metabolism.

Parsley (*Petroselinum crispum*)
Contains vitamins A, B, C and E, as well as calcium, phosphorus and large amounts of iron. Parsley is also a digestive tonic and useful in some cases of arthritis and kidney disease.

Watercress
Given daily to your cat, watercress is most beneficial. It tones the circulation and provides a good source of vitamins A, B, C, D, E and iron.

Alfalfa (Lucerne)
This is an excellent all-round supplement especially useful in debilitated animals. It stimulates both appetite and digestion, improving weight gain and overall condition. Give half a teaspoon of powdered alfalfa mixed with food each day.

Wheatgerm oil
A rich source of vitamin E (a natural antioxidant) and all the B vitamins.

Cod liver oil
Cod liver oil is a good natural source of vitamin A, necessary for good eyesight, and vitamin D, needed for bone growth and development.

Supplements for vegetarian cats
If you are considering feeding your cat a vegetarian diet, special supplementation will be necessary to meet all the cat's needs. Extra vitamins, specifically A, B_{12} and C are needed. It is even more vital to supply the amino acid taurine, normally only available from meat. Deficiency can lead to heart disease and blindness. Cats also require a preformed source of arachidonic acid, lack of which can lead to general poor health and reproductive problems.

Adding any of the above supplements to a vegetarian diet will go some way to preventing problems. Cod liver oil in particular contains large amounts of vitamin A and arachidonic acid, whilst milk and eggs can provide some of the taurine. However to ensure that sufficient is provided it is always best to give a specialized vegetarian cat supplement.

---

nate small amounts up to ten times a day. A cat's urine should be clear and yellow, although it has a pungent smell as it is quite concentrated. Any spots of blood in the urine, very frequent urination or straining by the cat to pass urine are the signal for an *immediate* visit to the vet as they can be symptoms of a life-threatening illness.

Adult cats defecate once or twice a day. The faeces should be firm, but not hard – or as a farmers' saying goes: 'the right consistency to pick up and throw over your neighbour's fence without making a mess of your hands!' High-protein foods cause a stronger smell, so if this is a problem you might want to try a change of diet.

A cat marks its territory with both faeces and urine. The more in evidence they are, the more territorial and assertive the cat. Most cats will scratch around their faeces, covering them with earth or litter but a dominant cat leaves them openly displayed to advertise its presence. In a household of several cats, one may cover up another cat's faeces to dull the olfactory message.

Urine, an even more common territorial

marker, is usually sprayed over a vertical surface. The cat backs up to a tree, bush, dustbin or fence post and lifts its tail. The tail will be held stiffly and may quiver at the tip, while urine is sprayed backwards in a powerful jet. Some cats can be seen repeating this process every few yards, using just a few drops of urine each time. Other cats will know that there is a feline in the neighbourhood and will gauge how long since it passed by the strength of the smell. Some will spray over the mark to impregnate it with their own scent. Contrary to popular belief, both female and male cats spray, although males do so more often, especially if they have not been neutered. Spraying usually only causes problems when it is done indoors (see p. 70).

Cats need a litter tray even if they have free access to outdoors. They do not always want to go outside, especially if it is rainy or cold, or if there is a bully-cat or a hostile neighbour in the area. They like privacy, so put the tray

CAT LITTER CHOICES

Calcium montomorillonite
A heavy, grey-coloured clay also known as fuller's earth. It is inexpensive, extremely absorbent – holding liquid up to 125 per cent of its own weight – and an excellent smell absorber. It also disperses in water, so small amounts can be flushed away. The disadvantages are its bulkiness and weight.

Attapulgite
A pure-white clay, which is lighter than calcium montomorillonite but less absorbent. It is usually disposed of in the dustbin, although small amounts could be carefully flushed.

Sepiolite
Another white clay, which can absorb moisture up to 110 per cent of its own weight. It is disposed of in the same way as attapulgite.

Moeler clay
This pinkish-brown litter is disposed of in the same way as attapulgite and sepiolite.

Sodium bentonite
A finely ground grey clay. It forms solid clumps when wet so that used parts of the litter can be removed easily without changing the whole tray. Although other clay litters form clumps when wet, sodium bentonite does appear to hold together better.

Wood-based litters
Softwoods that have been turned into sawdust pellets. They absorb up to 300 times their own volume of liquid and turn back into sawdust when wet. Their natural pine scents are released by moisture so they are good for liquid waste, but not so effective against the smell of solid waste (cats seem to have no objection to natural pine although they dislike pine disinfectants). These litters are completely biodegradable and can be flushed, burned or placed on the compost heap.

Washable, reusable cat litters
These are now a reality. One brand uses pelletized corncobs covered with food-grade paraffin wax. These granules come in a specially designed tray with slots at the bottom and another tray underneath. Urine passes through the granules and collects in the bottom tray, where the smell is trapped, so it need only be emptied every few days. This type of litter is not so effective with solid waste as it has no deodorizing properties, but a weekly rinse through will keep it fresh. As the litter itself does not have to be disposed of, it is the ideal choice for those living in an apartment.

Do-it-yourself litters
These range from sawdust, shredded paper, and earth to small, smooth stones that can be washed regularly.

New litter
The newest type of litter, available in the USA, changes colour when used according to the amount of acid in the cat's urine. This could be of benefit to owners of cats prone to urological problems.

in a quiet, out-of-the-way spot, maybe even in an easily-accessible cupboard. Covered trays are often popular as the cat can climb discreetly inside through a small doorway. One model incorporates a charcoal filter in the hood, which is said to cut down on smells.

Litter fillers come in various forms. The most common are clay – natural earths, which have been mined, ground and dried. Cats have their individual preferences. Some longhaired cats, for example, like the pellet-

Left. An anxious cat will draw in air through its mouth passing over the Jacobson's organ. By doing so, it can both smell and 'taste' the air.

Below. Spraying is usually carried out over vertical objects, such as the clump of grass shown here. Some cats carry out the motions of spraying without ejecting a drop.

ized softwood litters as they do not stick to the long fur on their legs. If you want to change the type of litter and your cat resists, place some of the old type on top of the new until it gets used to it. If your cat still absolutely refuses, there may be a good reason – it may be dustier, or contain something your cat does not like. There are deodorized litters on the market as well as separate deodorants that can be added to litter, but some cats refuse to use them and may even have an allergic reaction.

Cats may also refuse to use their tray if they dislike the smell of the cleanser used. Pine disinfectants are often the culprits. Use a household bleach, diluted in water, and thoroughly rinse and dry. A cat will often give itself a dustbath in a clean tray. Clay litters, in particular, are good at absorbing grease in the fur and acting as a dry shampoo.

Toxoplasmosis is an illness that causes much concern. Cats pick up the parasite *Toxoplasma gondii* from infected prey, from undercooked or raw meat or from raw, unwashed vegetables. Symptoms include weight loss, diarrhoea, pneumonia and fever, but even if there are no obvious symptoms, the faeces can be infective for up to five weeks. If the parasite is passed on to a pregnant woman, there is a risk of birth defect in her unborn child. There is, however, no need to panic. Approximately half the population of the western world is immune. One in two people has contracted toxoplasmosis at some time, usually without knowing it as it is often like a mild form of influenza, and the source was not cats, but undercooked meats (it is particularly prevalent among those who work as butchers). However, to be safe, pregnant women should let someone else clean out the cat's litter tray and disinfect it daily, meat should be thoroughly cooked and gloves worn if gardening. Another precaution

is to have your cat tested for antibodies to toxoplasmosis. If they are present in the blood and have not increased when tested again a week later, the cat has been infected previously and is now immune. Its faeces are not infective and it will pose no risk.

Roundworms inhabit the intestinal canal where they may prevent the proper absorption of food. They are cylindrical and can be 10 cm (4 in) long, although longer specimens have been found. Tapeworms, which live in the small and large intestines, have segmented bodies up to a metre long. Some of the segments may break off and can be seen wriggling around the cat's bottom like grains of brown rice.

Signs of roundworm or tapeworm infestation can include any of the following: diarrhoea, vomiting, constipation, weight loss, lack of interest in food, poor condition (a 'staring' coat) and pot bellies, especially in kittens. Roundworms are passed in the mother's milk to kittens or are eaten as eggs in the soil or litter tray. Tapeworm infestation occurs when a grooming cat swallows a flea with worm eggs in its system or by eating prey whose bodies contain encysted larvae. Specific worming treatments should be given for the particular type of worm involved as roundworm treatments will not kill tapeworms and vice versa. Cats need worming for roundworms every three or four months and for tapeworms every six months or so, especially if the cat is a hunter or has had fleas. Regular grooming also helps to keep your cat tapeworm-free. Cats can also suffer from lungworms, picking up the larvae when they ingest infected birds and rodents. Lungworms rarely cause problems, often nothing more than a mild cough.

Other less common parasites are hookworms, which are bloodsuckers that can damage the lining of the intestine. Symptoms are diarrhoea, loss of weight and poor coat condition. Whipworms are found in the large intestine and threadworms in the small intestine. These are also bloodsuckers and can cause anaemia. Flukes, found in the small intestine and pancreas, are carried by a fish and can cause diarrhoea, anaemia and jaundice.

One end of the cat can cause just as many problems as the other but you will minimize them if you make an informed choice of cat food, provide a litter tray, pay attention to basic hygiene and worm your cat regularly.

WORMING REMEDIES

Read 'Before You Start', pp. 16-17.

Recurrent worm problems can indicate general ill health or poor resistance in both kittens and adults. Diet can be important and fasting may be necessary where it proves difficult to eliminate worm problems effectively. Unfortunately, natural worming remedies can be unreliable. In cases of roundworm, conventional treatment is often recommended, particularly for kittens, which are often infested in large numbers through their mother's milk.

HOMOEOPATHIC REMEDIES

Cina (Wormseed)
Suits irritable and angry cats and can be used to help eliminate roundworms. A guiding symptom is intense itching of the anus.
SUGGESTED DOSE: Cina 30, twice daily for 7 days.

Abrotanum (Southernwood)
This is a roundworm remedy. It is useful in cases where weight loss is evident, despite the fact that the cat has a good appetite. Digestion is poor and constipation may alternate with diarrhoea, the stools containing undigested food. Another symptom suggesting Abrotanum is a distended abdomen.
SUGGESTED DOSE: Abrotanum 6x, twice daily for 10 days.

Teucrium marum (Cat-thyme)
Itching of the anus suggests this remedy, which can be used to treat roundworm infestations.
SUGGESTED DOSE: Teucrium marum 6, twice daily for 10 days.

Filix mas (Male fern)
Filix is a tapeworm remedy, notably where constipation is a feature.
SUGGESTED DOSE: Filix mas 3x, twice daily for 10 days.

Granatum (Pomegranate)
This is a useful remedy for tapeworm infestations in cases where there is weight loss despite eating well. Itchy paws are a guiding symptom.
SUGGESTED DOSE: Granatum 3x, twice daily for 10 days.

HERBAL REMEDIES

INTERNAL
Garlic (Allium sativum)
Amongst its many properties, Garlic has an anti-parasitic action. In order to eliminate worm burdens, however, large amounts would have to be administered to the cat. It is generally better, therefore, to give smaller daily doses as a preventative measure.

A number of other herbal remedies such as Pumpkin Seeds, Cucumber Seeds, Cayenne Pepper, Rue and Wormwood can be used to eliminate worms in conjunction with a fasting regime. This is, however, best carried out under supervision and is not always a practical method where cats are concerned.

CHAPTER TWO

Exercise, Rest and Play

Play is serious business for a kitten. As soon as it can stagger around on its four wobbly legs at about three weeks, it will spend much of its waking hours tumbling over, biting its littermates and chasing its mother's tail. These are no idle games, but important lessons in the skills it may need to survive as an adult. Play fights prepare a kitten for the real thing, when it may have to defend its territory or its mate, and tail-chasing is the first step towards hunting.

Nowadays, few cats need to hunt for food and many would be hard put to catch anything even if they wanted to. Urban environments and small territories mean that many cats do not take as much natural exercise as the free-ranging rural cat, which may roam a territory as large as 100 hectares (250 acres). Those kept constantly indoors for their own safety are even more severely restricted.

Kittens are naturally active but adult cats, given little opportunity for exercise, may become lazy, overweight and lethargic.

Left. The fixed look and stealthy prowl of the hunting cat. The hunting instinct is triggered by movement, not hunger, explaining why cats will kill prey they do not eat.

Moreover, few owners who attend exercise classes to keep themselves fit consider devising exercises for their pets! Yet play keeps cats healthier. It tones their muscles, strengthens their cardiovascular system, reduces stress, which can cause behavioural problems, and alleviates boredom, which can lead to destructive behaviour. Regular play sessions can be fun for both owner and cat and it strengthens the bond between them.

Ten to twenty minutes of play each day will keep your cat fit and happy. Longer than that and you may find its attention wanders, although some will play for hours if allowed. Just as in an aerobics class, have a gentle warm-up before beginning the exercise session and a cool-down period at the end. Try to keep to the same time each day – cats are creatures of habit – and do not exercise your cat within an hour of a meal.

Start by rolling a ball for your cat to chase. Remember that it is 'hunting', so will only chase the ball if it is rolled *away*, as a prey animal would run away from a cat. Most cats will simply look perplexed if a ball is rolled directly towards them. Cats adore anything that rolls erratically, so collect a toy-box of safe playthings such as egg-shaped plastic

Below. *Here the kitten is 'fighting' its mouse as if it were another kitten. In a real fight, the feline underneath would draw up its back feet to claw at the belly of the other cat.*

containers, empty plastic 'lemons', walnuts, pine cones, wine corks (champagne corks are particularly good fun), cotton reels and disposable drinking cups.

Make your own catnip 'mice' by cutting scraps of any tough fabric into tubes 10-20 cm (4-8 in) long and stuffing them loosely with catnip so they are soft and floppy. They do not have to be made of fur fabric or even to look like mice, as your cat may never have seen a mouse and will enjoy playing with anything it can toss around. If you buy a ready-made toy, try to pull off ears, eyes, tails, whiskers, legs or any other loose appendages that could get stuck in your cat's throat. In some countries such as the UK there are no safety standards laid down for pet toys.

If you throw a twisted piece of paper, some cats will 'fetch' it. Other bright cats will join in games of table tennis, sitting on the table and batting back the ball with their paws.

Below. *This kitten is instinctively giving its toy mouse a 'killing bite'. Kittens' canine teeth are ideally spaced to sever the spinal cord of a mouse, while those of an adult cat can kill a rat.*

Some will play football with a small, light ball and in the USA there is even a report of a cat which plays basketball on its own scaled-down court!

You can watch your cat hunt without leaving your chair by making a simple fishing-rod. Tie some stout cord to a pole of about 1 m (3 ft) in length and attach either a small toy or a scrap of strong fabric such as denim to the other end of the cord. Then sit back and dangle the lure in front of your cat. It will have enormous fun sneaking up on the lure and pouncing – it may be the only hunting opportunity many cats get. Outdoor cats, even if they do not hunt live prey, can often be seen chasing wind-blown paper scraps, or batting at waving plants and grasses. Indoor cats rely on their owners for hunting play.

All cats love hunting, whether they stand a chance of catching anything or not, and show enormous patience when their hunting instinct has been aroused by something moving. The cat crouches low for minutes on end, its body frozen to the spot (except perhaps for a twitching tail) and its eyes transfixed on its prey. When it judges the time is right, it begins to stalk, either edging or running towards the prey, with belly close to the ground, elbows high and tail flat out behind and twitching. At several metres the cat hides behind the nearest cover – which may not mask it completely – while it decides whether to pounce or to stalk still nearer. If it decides to pounce, it begins to 'wind up' its hind legs, pumping them up and down, to generate enough energy for a spring. The tail twitches more furiously until the whole rear end moves from side to side. Then in an instant it bounds forwards. If the prey runs and tries to dodge, the cat can pivot and turn very swiftly. Its tail becomes a counterweight, swinging to the opposite side from the turn. Once the cat lands on the prey with its fore-

This kitten is brushing up its hunting technique by chasing bees. Although such a hunt will keep it active, amused and fit, it could end up with a badly-stung mouth or nose!

paws it can make a killing bite, grasping the neck and biting through the spinal cord with its teeth.

If the cat fails to make a clean killing bite, it may be forced to act defensively. A bite from a rat or even a mouse could prove lethal to a wild-living cat, so the cat's instinct is to preserve itself from damage while continuing the hunt. It does this by batting at the head of the prey, to make it lower its head defensively. From this position, it is unable to leap up and bite the cat's vulnerable face or neck area. Once it manoeuvres the prey into the proper position for a killing bite, the cat will

THE HUNTING INSTINCT

Many owners have unrealistic expectations where cats and hunting are concerned. Cats are creatures of instinct. Their hunting instinct is triggered by movement, as you will know if you have ever watched a cat 'hunting' a blade of grass waving in the wind. Equally, a cat will hunt a rat, a mouse or a pretty songbird. It cannot tell the difference and does not appreciate that humans generally detest rats and treasure songbirds. To a cat, they are all moving targets and potential food. This instinct will save a wild cat from starvation. A domestic cat does not lose this instinct as it never really knows that you will fill its bowl at mealtimes. Do not praise your cat for killing a disease-carrying rat and shout at it for killing a sparrow, or you will end up with a very confused feline.

A movement in the grass attracts the cat's attention. A slow and careful reconnoitre is now carried out while the cat decides if the prey is within hunting distance.

Left. *The cat makes itself as small as possible while it stealthily approaches its prey, eyes fixed and ears alert.*

The cat has decided that the prey is within range. It hides behind any available cover, even a blade of grass! If necessary, it will move forward to find more cover.

Within leaping distance, the cat 'winds up' for the pounce. The cat's rump will move from side to side, hind legs may move up and down, while the eyes never leave the target.

When the cat judges the moment is right, it leaps. The paws will hold down the prey animal while the teeth deliver a fatal bite to the spinal cord.

Below. *Kittens first learn hunting skills by chasing their mother's tail. Later, a hunting female will bring dead prey, probably mice, back to the nest. The kittens may do no more than lick the prey but later will learn to eat it. Kittens that do not learn to eat prey at this stage may never know what to do with it and fail to recognize prey animals as food.*

Above. *Although cats are popularly supposed to hunt alone, these two silver tabbies belie the image of the cat as a solitary hunter. Both cats are ready to pounce and it looks as if the cat on the right can already taste its next meal.*

Right. *A cat will try to find a quiet, undisturbed place to eat what it has caught. This may mean bringing it back to its home or at least to its own garden. Although many cats catch shrews, like this one has done, not many will eat them. They are said to taste bitter.*

dispatch it swiftly. This batting movement is often misinterpreted as cruelty on the part of the cat, when in fact it is a simple act of self-defence. Tossing prey in the air also has a purpose – it stuns or disorientates the animal if it is alive, and confirms when it is dead. After the excitement of the chase, if the prey animal stops moving, the cat will leave it to wind down. It will often walk some distance away, returning to carry its prey to a safe eating place. If, in the meantime, the prey begins to move, the movement triggers the hunting sequence again.

Cats often bring their prey home, not so much as a gift for their owners, but for somewhere safe to eat. Many cats will growl at their owners if they try to remove the prey. If it is a large bird, the cat will begin to pluck it, then turn its head to the side and spit the feathers out. After a few mouthfuls, it licks its own fur to clean off any feathers sticking to the barbs on its tongue. Small birds are consumed feathers and all, and rodents are eaten whole. If the cat is not hungry, it will try to hide the body in a safe place, to eat later. In the wild, it would scratch earth and leaves over the dead prey. A domestic-living cat might even leave it in its feeding bowl.

Owners who are upset when their cat hunts need to remember that cats do not value birds above rats as most humans do. If you are a cat *and* a bird lover you could seek advice from your local bird charity on making a cat-proof feeding table.

Although a cat that hunts regularly may be supplementing its diet with roughage from fur and feather, calcium from bone and vitamins from the contents of the prey's stomach, there are dangers. It may develop tapeworms, which spend part of their life cycle in prey animals, or eat a poisoned rat or mouse, which could make them ill or even kill them.

The skills of self-defence are learned in kit-

Kittens learn their hunting skills by chasing their mother's tail. Mums will obligingly wave their tail from side to side. In this case, the waving tail probably indicates annoyance!

tenhood, along with hunting behaviour. Two cats about to fight will first stare at one another. This is a threatening gesture in the cat world. An individual who has learned that it is not a good fighter from play or from battle with its contemporaries can simply look away and the fight will not happen. It may turn its back on the aggressor to show that it is unwilling to fight, and it is very rare for a cat to attack if its opponent has its back turned. If both cats decide they are evenly matched and refuse to give way, they both put on a show of strength. This involves hissing, arching their backs, raising their fur and fanning out the fur on their tails until it looks like a bottle-brush. Ears back, they approach one another, almost on tiptoe, tails thrashing, emitting threatening noises until one aims a bite at the other's neck. One cat will throw itself on to its back – a good fighting position as it leaves teeth and four sets of claws free to fight. The cat on top rakes as best it can at the other's face and belly, biting and slashing while trying to remain out of reach. The pair then jump up and face one another again before repeating the entire sequence. This continues until one cat, injured or tired, lies on the ground in a submissive position. The other cat then saunters off victorious, its point made. It inflicts no more injury once the vanquished cat gives up.

Cats retain all their old instincts when living in a domestic situation. If you scratch a cat's belly, for example, this sets off an association with the scrabbling of an attacking cat and your pet will turn swiftly and bite you. Many cats do seem to be apologetic afterwards, however!

They also learn new skills from their human companions. A domestic cat is, for example, more vocal than a feral cat, simply because humans talk to it. Few owners feed their cat in silence. They ask if it is hungry, or

Adult cats, whatever their age, still love to play. If not given toys, they will make their own, pouncing on waving grasses or plant tendrils, or chasing scraps of paper as they blow in the wind.

tell it that it is several hours to the next mealtime. Cats quickly pick this up, so when they next go to their feeding bowls they 'converse' in the same way. Domestic cats also realize how important the sense of touch is to us and learn to communicate what they want in a way we understand. If an owner stops stroking a cat, for example, it will sometimes stretch out a paw to touch the owner's hand or arm to ask them to continue. Some cats, if ignored while their owner talks to someone else, will even place their paws over their owner's mouth.

PREVENTATIVE HEALTH CARE

Read 'Before You Start', pp. 16-17.

There are some herbs which can be given safely on a daily basis, to help prevent problems occuring. It is better to think of these as tonics rather than medicines in this respect.

Garlic *(Allium sativum)*
Garlic is commonly referred to as 'nature's antiseptic' as it has anti-bacterial, anti-viral, anti-fungal and anti-parasitic properties. It is particularly good at preventing respiratory problems and is also beneficial in digestive disorders, where it supports the growth of normal gut bacteria.

Nettles *(Urtica dioica)*
Nettles provide support for the body, strengthening the whole system. They are of particular value where skin problems are present.

Burdock *(Arctium lappa)*
This herb restores the body to a state of balance and helps maintain it in a state of equilibrium. It cleanses the system of impurities, stimulates the digestion and appetite and can alleviate some skin conditions.

Hawthorn *(Crataegus oxyacanthoides)*
The berries are an excellent tonic for the heart and circulation, improving blood flow through all the major organs. Use a liquid homoeopathic preparation, containing the 1x potency. It is often combined with the Night Blooming Cereus *(Cactus grandiflorus)*. Give 2 or 3 drops of this tonic to your cat 3 times daily in its food.

Buchu *(Agathosma betulina)*
Given to your cat every day, Buchu can prevent urinary problems such as cystitis or urethritis occuring. Additionally, it can also help to stop gravel from forming in the cat's bladder.

Oats *(Avena sativa)*
Oats have a tonic effect on the whole nervous system and can provide supportive treatment in cases of epilepsy. Give a little each day as porridge mixed with the food. A few drops of the homoeopathic mother tincture given twice daily can help in cases of general debility.

Celery seed *(Apium graveolens)*
Celery seed can alleviate problems with rheumatism and arthritis, acting as a tonic for both muscles and joints. It is also a urinary antiseptic, useful in preventing cystitis and associated problems.

Enriching the Environment

Most domestic cats lead unnatural lives in one way or another. Many are left alone without entertainment or any significant territory of their own. Others are kept entirely indoors. Statistics compiled by the Pet Food Manufacturers' Association show that most cat owners are busy, young, urban professionals who believe that cats fit in with their life-style better than other pets. Not surprisingly, cat ownership is growing faster than dog ownership. Some owners keep the cats indoors to protect them from diseases, from car accidents or from attack by other cats, dogs or unfriendly people.

Cats can live happily indoors, but only if the owner meets all their needs. One of the essentials is companionship — cats are *not* solitary creatures, they grow up in litters with a mother cat to boss them around. Feral cats congregate in groups and even house cats find feline (or human) friends locally. In the home, we become the 'chief cat' and our cat looks to us for guidance and companionship.

All cats enjoy a compatible feline friend, but for an indoor cat with a working owner, it is probably essential. If both cats are neu-

tered the sex and age difference is relatively unimportant, but temperament does matter. An assertive cat is not a good companion for a shy one. I also find that unrelated cats usually get on better than siblings or mother and off-spring once they are adults.

Indoor cats must be fed a well-balanced diet as they are unable to supplement any deficiencies by scavenging or hunting. They also rely on their owners to provide green-stuff – to aid digestion and to help eliminate hairballs. If there is no grass to nibble, they will eat your houseplants (they may do this anyway). To ensure a regular supply, plant several small tubs with seed at 10-day inter-vals. Many pet stores or cat-equipment sup-pliers sell cocksfoot grass seed but, if you cannot get it, any ordinary grass seed will do as long as it has not been treated with chemi-cals. It is also important to make water avail-able at all times.

A scratching post is also essential. Strop-ping hones a cat's claws, helping to shed the worn claw sheath, and the stretching and scratching motions exercise its muscles. It also imprints the object with scent from pawpads and so gives the cat a sense of security. The post must be tall – 1 m (3 ft) is ideal – so the cat can stretch fully.

It is in a cat's nature to climb and hide, and the higher its resting-place, the more secure it will feel. A large, old-fashioned house with plenty of stairs, large furniture and nooks and crannies is more suitable than a small, sparse, modern apartment. However, wooden and plastic climbing frames are available with a carpet or rope covering and shelves at different heights. Some include

Cats need exercise as much as people. Although kittens are naturally active, adult cats living in a domestic environment need to be exercised. It tones the muscles, strengthens the cardiovascular system, reduces stress and relieves boredom.

boxes at the base with entrance holes, or tubes halfway up for hiding and ambushing. Any design will provide wonderful exercise for a housebound cat, helping to keep its muscles in trim and its stomach slim.

Light and sunshine are also vital. Cats love the warmth of a sunny window-sill, especially if the view is interesting, and the sunshine synthesizes vitamin D on their coats, which they then lick off. Unfortunately, these rays cannot penetrate glass so some owners keep a window partly open and fit a wire screen in front to prevent their cat falling out. Cats *do* fall out of windows. They may roll off a sill when asleep or, in the thrill of the chase, follow an insect through an open window.

Right. *Climbing is excellent exercise for all a cat's muscles. Trees have the added bonus of being superb natural scratching posts.*

Below. *Cats prefer stropping on vertical surfaces but will also strop horizontally if the surface is particularly tempting. Smooth fabrics are less likely to be stropped than textured ones and plastic will not be stropped at all.*

Finally, a cat needs to play. If you are not home to play with your cat, leave toys for it to play with alone. Bring out one or two each day, leaving the others shut away so that it does not become bored with its toys. Balls are a favourite but are easily lost under furniture, so leave them in a cardboard box. A shoe-box with a lid and paw-sized holes cut in it can hold either a ball or a few treats for your cat to hook out and eat. Cotton spools, threaded together, and securely tied to a door handle are excellent for batting around. Most cats enjoy catnip toys as the active chemical, nepetalactone, in this mint-like herb both relaxes and excites. When not in use, keep them stored in an airtight container to preserve essential oils and scent. Some cats are easily pleased when it comes to toys, enjoying something as simple as a folded newspaper left on the floor and will spend delighted hours shredding and treading it.

It is dangerous to leave loose string, cord, wool or thread for your cat to play with. Once a cat gets the end in its mouth, it is unable to spit it out because of the barbs on its tongue and must swallow it. Surgery may be needed to remove it. Never leave threaded needles around either because if the cat begins to swallow the thread, it swallows the needle too. Otherwise treat your indoor cat like an inquisitive toddler, for it is just as likely to get into danger. Keep sharp objects shut away and do not leave anything cooking when you are not around to supervise. Remember that cats can climb the most unscalable objects and can squeeze through tiny gaps just 5 cm (2 in) wide.

The indoor/outdoor cat benefits enormously from an 'indoor cat' environment too. In fact, made as comfortable as this, the indoor/outdoor cat may spend as little as five or ten minutes a day outside.

There are two types of indoor/outdoor living. The first is where a cat is given a cat-flap to come and go when it pleases. The second, and preferable option, is when the owner lets the cat in and out on demand. The advantage of this is that the owner knows whether the cat is indoors or out at any given time and will notice sooner if the cat has been gone an unusually long time.

Many cats supplied with a cat-flap show a marked reluctance to use it and continue to demand that their owners open doors for them. There are several reasons for this. Most cat-flaps have to be pushed open with the head, and some are extremely stiff. Few owners would ever go outdoors if that was the only way they could open their doors! Compare flaps in the store by pushing gently.

Cats like to see what is outside before venturing forth. With a cat flap, once they are partly through they are committed and cannot change their minds if a dog or another cat is waiting outside for them. Cats who go through an open door will look outside before stepping out. So see-through plastic cat-flap doors are better than opaque ones as they give a cat some view of the outdoors.

Some cats refuse to use flaps because they have been trapped in them in the past. Sometimes a cat pushes a paw through the flap then changes its mind and tries to withdraw, leaving the paw trapped. The harder the cat pulls, the tighter and more painfully it is caught. Check this when buying your flap — push a finger through on the threshold of the flap then try to withdraw it. If it gets caught, so could your cat's paw. Side-opening flaps seem to cause fewer problems in this way than top-opening varieties but the soft, rubber ones are better still.

Right. *This careful pat of the ball is to ensure it is 'dead'. A live prey animal would be tossed into the air to stun it.*

Some types of flap open automatically at your cat's approach if it wears an appropriate collar. One type has an electronic device on the collar which transmits a signal that opens only the flap matching its frequency. The disadvantage is that some bright felines have been known to follow swiftly behind a collar-wearing cat, so that two cats enter for the price of one. The intruder may then be trapped indoors, until a human lets it out. There is also an electro-magnetic collar, which is not selective and allows the cat to enter any home with such a flap. It is, however, possible for pieces of wire or metal to become attached to the magnet on the collar, potentially causing injury if they catch on the cat's skin. Although electronic flaps have disadvantages, you are less likely to have unwelcome feline visitors than with other types. With a basic flap, you cannot put off visitors without discouraging your own cat.

If you do install a cat-flap, make sure that it is more than arms-length from any lock as burglars have been known to put their arms through and unlock doors. Do keep it locked at night, too, with your cat on the inside, as more cats are killed, injured or lost at night than during the day. If it seems less trouble to open the door to let your cat in and out, you could be right!

Some cats live entirely outdoors but in a man-made environment. These are usually the stud cats — pedigree males used for breeding — or neutered pets with 'anti-social' habits, such as spraying. These animals are much-loved but their life-style is not accepted in the home. It is a myth that stud cats do not live as part of the family because they are vicious. Well-cared-for stud cats are as loving as any other; it is simply that the male hormone, testosterone, induces them to spray as an invitation to females. Most owners, understandably, cannot live with the smell.

Their outdoor enclosures are usually comfortable but not luxurious. The minimum size is about 1 m (3 ft) by 2 m (6 ft) and includes an outdoor run, concreted over for hygiene, with an indoor chalet for comfort. The chalet includes a comfortable and draught-free bed and a heater for cold nights. Owners with sufficient grounds build their run around growing trees so that their cat has something to climb up and strop. Otherwise, scratchers and climbers are necessary. Even

more necessary is stimulation and company. Most outdoor runs are built in an area where there is something for the resident cat to see, as near as possible to the owner's home and preferably with other cats within eye-shot. Owners spend hours each day inside their cat's run, reading, writing letters, snoozing and generally keeping their cat company. Cats living in runs without human company can become withdrawn, depressed or even aggressive.

The last alternative is the completely outdoor cat. Rare today, the outdoor cat is found mainly among farming communities. It is sometimes left to fend for itself, sometimes fed, sometimes given fresh milk to drink, but expected to earn its keep by killing rodents. Such cats are rarely pets, rarely friendly, and either remarkably healthy or prematurely deceased. Independent studies conducted by

This sleek feral cat looks healthy and contented, but its life span could be less than half that of the pampered, indoor cat. Disease is rife in feral colonies and food often in short supply.

several US vets have estimated that the life-span of a completely outdoor cat is about 6 years, compared to 8 to 10 years for the average indoor/outdoor cat. The completely indoor cat averages 12 to 15 years — sound ammunition against those who believe keeping cats indoors is 'cruel'.

Whatever a cat's environment, one thing is certain. It will sleep for about 16 hours each day — or longer as a kitten — making sleep just about the most important part of a cat's day. It is an ideal way of conserving energy for those short, sharp bursts of intense action during hunting or self-defence. The pampered house cat may not have to fend for

Cats are not being lazy when they sleep for as much as 16 hours a day – they are conserving energy. Hunting down that can of food two or three times a day can be totally exhausting!

itself, but it stocks up on sleep regardless. Another way cats conserve vital energy is to rest in a warm place, so that the body does not burn up calories keeping itself warm. All cats are adept at finding warm spots. Some sleeping places may seem strange until you discover a central-heating pipe runs underneath. Some are obvious but unwelcome; it is not easy taking clean sheets out of an airing cupboard when a protesting 5-kg (11-lb) cat is sleeping on top. Some are downright dangerous; most owners have at some time detected the smell of burned fur as their cat sleeps on oblivious by the fireside (cat's fur is such a good insulator that the intense tem-

perature does not reach the skin).

You can persuade your cat to sleep where you want by choosing the right bed. It needs to be warm, draught-proof and large enough to contain the cat when it stretches right out – and all cats stretch when hot. When choosing a bed for a kitten, either find a large one it can grow into, or use one that can be replaced. Such a bed need not cost a lot – a stout cardboard carton is ideal. It can also be replaced regularly and the bedding thrown away or washed to discourage fleas and other parasites. Choose a carton without a strong smell – one which contained detergent, for example, will not appeal to your cat. You can make it cosy with layers of newspapers on the bottom and an old blanket on top. Alternatively, buy one of the many artificial fur sheets sold as pet bedding, which are washable and very comfortable.

SLEEPING

Humans sleep for approximately one third of their lives. Cats sleep for two thirds. Kittens sleep even longer than the average 16 hours of the adult cat because they use up so much energy in play, while learning the skills of hunting and defence that they will need in adulthood. New-born kittens will sleep almost all the time that they are not feeding and some seem to be able to suckle and sleep at the same time.

When a cat first falls asleep, it will not be totally relaxed and may sleep sitting up or with the head held high. At this stage, a slight noise will wake it. After about half an hour, rapid-eye-movement (REM) sleep begins. This appears to coincide with the dreaming state in deep sleep. The eyes, although closed, move from side to side, the limbs may twitch or the cat grimace, depending on its dream. Cats are believed to spend up to 60 per cent of sleeping time in REM sleep, three times more than humans.

Below. Cats spread themselves out when the temperature is high. When buying a bed it is wise to allow for this as a cat will not use a bed that is too short to allow it to stretch. Beds for kittens will soon become too small for an adult cat.

FAVOURITE PLACES

A cat's favourite sleeping places are always warm and often high. The higher a cat can rest, the more secure it feels. It cannot be easily attacked and it has a vantage point from which to survey everything that approaches. In a group of cats, domestic or feral, the most dominant cat can be spotted immediately as it will be the one with the highest sleeping perch.

In new surroundings, a cat will demonstrate its insecurity by refusing to come down from the top of a wardrobe (if it has not decided to hide under the bed instead). A nervous cat will appreciate the security of a bed placed in a high situation and being left in peace there.

Most domestic cats will try to combine warmth with height by sleeping in airing cupboards or on top of central-heating boilers. Some farm cats have taken over abandoned birds' nests, even choosing to give birth in them, while other cats choose precarious high perches on stair handrails or even on top of doors. Our pet cats demonstrate a high degree of trust and love for us when they deign to sleep on the floor.

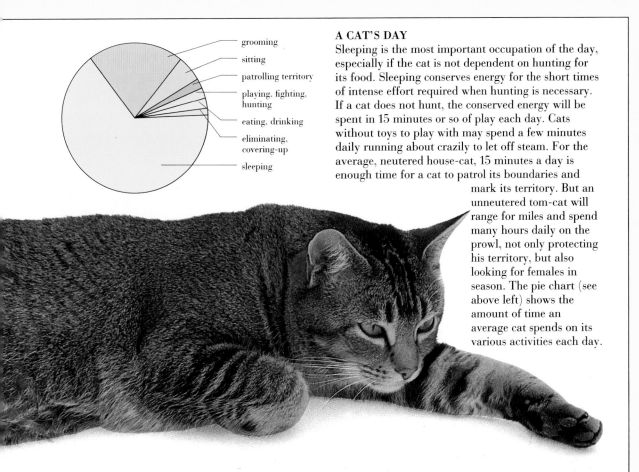

grooming

sitting

patrolling territory

playing, fighting, hunting

eating, drinking

eliminating, covering-up

sleeping

A CAT'S DAY

Sleeping is the most important occupation of the day, especially if the cat is not dependent on hunting for its food. Sleeping conserves energy for the short times of intense effort required when hunting is necessary. If a cat does not hunt, the conserved energy will be spent in 15 minutes or so of play each day. Cats without toys to play with may spend a few minutes daily running about crazily to let off steam. For the average, neutered house-cat, 15 minutes a day is enough time for a cat to patrol its boundaries and mark its territory. But an unneutered tom-cat will range for miles and spend many hours daily on the prowl, not only protecting his territory, but also looking for females in season. The pie chart (see above left) shows the amount of time an average cat spends on its various activities each day.

TEMPERATURE

You can always tell the temperature by looking at your cat. Cats spread themselves out (see above) as it gets warmer. By this cat's posture, the temperature must be well over 75°F. If it curls its hind legs under its body, the temperature will have dropped by at least 5°F. At around 65°F the front paws will curl under the head; at 60°F the head will lower and the tail begin to come around the body (see left); at 55°F the tail is over the nose; and at any lower temperature the cat is curled into a tight ball, tail covering nose and ear in an attempt to keep the rapidly cooling extremities warm.

If you prefer to buy a ready-made bed, there are hundreds of designs to choose from, from a regal four-poster to a humble bean-bag. Modern beds are made in washable fabrics in a range to suit any decor. If your home is particularly draughty, your cat may like a hooded, igloo-style bed. It may treat it a little warily at first, but find that the dark interior is an ideal place from which to ambush companion cats.

Beanbags are becoming very popular as pet beds because the polystyrene bead fillings do not harbour fleas. They are machine washable, although they take time to dry and need a good shake. Do ensure that the beads are fire-retardant as cat beds often finish up next to heat sources. One disadvantage of beanbags is that some cats are confused by the sound and texture and mistake them for

Cats always try to find a warm place in which to sleep. Many have their favourite sun-trap in a garden. Make sure that your cat's bed is warm enough and free from draughts.

litter trays. If this happens, all you can do is to keep washing the bag to rid it of the scent and, if the behaviour continues, throw it out and use another type.

You can also buy bed heaters for pets. Like electric blankets only rigid, these fit under the bedding and stay warm when plugged in, while using little electricity. A warm hot-water bottle covered by a blanket will also warm up a bed for an old or sick cat or kitten but will only stay warm for a few hours.

Having provided a warm and comfortable sleeping place, let your cat spend its nights in comfort indoors. The idea of 'putting the cat out' is an old-fashioned and dangerous notion. Cats are more likely to be killed, injured or lost during the hours of darkness than during the daytime. A cat brought up from kittenhood to spend its nights indoors will not even ask to go outside at night. Wild-living cats may have to hunt at night because their prey is nocturnal but domestic cats are happily diurnal like their owners. If your cat has been accustomed to spending nights outdoors with a previous owner, close it into a convenient room with food (it will sleep better with a full stomach), water, a litter tray and a comfortable warm bed, then ignore any noise it makes (which may be considerable). Within a night or two it will begin to understand that the night is for sleeping, at least while you are chief cat!

REMEDIES FOR STRESSFUL SITUATIONS

Read 'Before You Start', pp. 16-17.

Placing a cat in a basket and then going on a car journey can be a frightening experience. Some cats are completely terrified, working themselves into a state of hysteria. Others will salivate profusely, sometimes vomiting or passing stools in the basket. A day at a cat show can also be a worrying experience.
(See Carrying Your Cat, pp. 115-116.)

HOMOEOPATHIC REMEDIES

Aconite (Monkshood)
Aconite will help alleviate fear and anxiety. It is useful for cats which are nervous at shows and for those frightened by car journeys.
SUGGESTED DOSE: Aconite 1M, 1 dose the night before. Repeat in the morning and again shortly before the full effect is needed.

Gelsemium (Yellow jasmine)
Fear is also a feature of Gelsemium, together with a desire to be left alone, a degree of apathy and sometimes restlessness. The cat that suits this remedy wants to hide away. Anxiety and stress of any cause which results in involuntary urination or diarrhoea may respond.
SUGGESTED DOSE: Gelsemium 30, twice daily, starting 24 hours before travelling or showing.

Argentum nitricum (Silver nitrate)
In contrast to Gelsemium, this remedy is associated with tense nervous agitation. Restlessness and excitement are also present. Involuntary stool and urination, fear of crowds or claustrophobia are useful indicators.
SUGGESTED DOSE: As for Gelsemium.

Cocculus (Indian cockle), Petroleum (Crude rock oil) and Tabacum (Tobacco)
These 3 remedies can be used to help travel sickness and prevent vomiting. Salivation is prominent with Petroleum, but trial and error is needed to find which works the best.

SUGGESTED DOSE: 2 or 3 doses, 30 minutes apart, prior to travelling.

BACH FLOWER REMEDIES

Rock rose (*Helianthemum nummularium*)
Rock rose will calm a frightened cat and help allay terror and panic. Dose with 4 drops every 15 minutes until the cat has calmed down.

Mimulus (*Mimulus guttatus*)
The fear indicated by Mimulus is less intense than that of Rock rose and is always caused by things known to the cat, such as a journey in a basket. To calm your cat in such cases, give 4 drops of prepared remedy, 4 times daily, starting the day before if possible.

CHAPTER THREE

Making Friends and Families

For centuries the cat has been considered a mysterious creature – aloof, unapproachable, unfathomable. Our understanding of cats is inevitably partial, but we can learn a lot just by listening and watching – learning to interpret their body language.

One of the most expressive parts of a cat is its tail. A fighting cat holds its tail up like a brush, a sleepy cat curls its tail over its nose to cut down oxygen intake and hasten sleep, and a relaxed and happy cat holds its tail straight out behind, slightly dipped. Once it sees something of interest, a cat may twitch its tail upwards a couple of times. If it is indecisive, it will twitch it from side to side, if angry or about to pounce, it will thrash it sideways.

When greeting you, a cat will hold its tail high and maybe bent over at the tip. It may bounce upwards, standing on its hind legs, in an attempt to reach and greet you. Afterwards it may turn its rear towards you as an invitation to sniff! It may follow up the wel-come by rubbing its chin around your legs or hands to mark you with its scent – the main scent glands are around the chin and at the base of the tail. When cats rub their chins on posts or fences they are impregnating them with their scent.

A happy cat will hold its head and ears high. When extremely happy – while eating, for example – the ears may be swivelled slightly to the side. Ears swivelled back mean a cat is contemplating attack, while a defensive cat will flatten its ears sideways to save them from the other cat's claws.

A stare is considered threatening behaviour in the cat world – fighting cats stare hard at each other. Often a fight can be averted simply by placing something, a newspaper for example, in the cats' line of vision. Unable to see one another, they wander off and do not bother to fight. This also explains why cats seem to be attracted to people who dislike cats. Someone with a cat phobia will avoid looking at a cat who interprets this as a lack of threatening intentions. If instead they stare at a cat entering a room, it will have nothing to do with them. (Ailurophobes rarely take this advice because secretly they are flattered by the attention!)

Left. This alert cat certainly has its eye on something. If we watch cats as carefully as they watch everything around them, we will learn to understand them much better.

Right. Eyes half-closed in contentment, tail curled ecstatically, this cat is welcoming a much-loved owner home. By jumping on the window-sill, it can rub against its owner, sharing scents. A cat that cannot jump onto something when greeting a human will bounce up on its back paws to make itself as tall as possible.

Below. Cats can detect sounds inaudible even to dogs and are also better at locating the source. Their ears are designed to detect the high-pitched sounds of their prey. Lower-pitched sounds, however, are more of a problem for the cat to distinguish.

So to make friends with a cat, look at it sideways without staring until it is used to you. A nervous cat will be reassured if you blink slowly at it and yawn – both signs of friendly intentions. A cat that keeps its eyes half-closed is displaying contentment. Pupils enlarge if the cat is alert and interested in what it sees. If you blink at your cat and it blinks back, this is the feline equivalent of a kiss.

An anxious cat twitches its ears and licks its lips rapidly. When very worried or nervous, it will flehm, gasping and inhaling air through its open mouth, to taste and smell the air for danger.

Whiskers also help to show how your cat is feeling. When fighting, they are drawn back against the sides of the mouth to emphasize the snarl, but if a cat is anticipating something pleasant they will be pushed forward so far they almost meet. Another sign of pleasure is kneading or paddling, when a cat uses its front paws to push rhythmically up and down, purring all the while and possibly dribbling. The origin of this behaviour goes back to kittenhood. The cat's first pleasurable experience is feeding from its mother while it paddles up and down with its paws to encourage the milk flow and its mother purrs. In adult life, the cat will repeat this movement and purr when very happy. The rougher the purr, the happier the cat, unless a respiratory illness has left it with a permanently rough and rasping purr.

The opposite of the purr is the hiss, which is designed to deter by its noise and by expelling air in the opponent's face. If a cat sniffs your face in the act of greeting, sniff back. Try not to blow or breathe out heavily, as the cat may take this as a silent hiss. This may well be why cats hate to be laughed at – when laughing we show our teeth and expel air forcibly – just like a hiss.

Some cats growl to show displeasure although not all seem able to do so. It appears to be a learned behaviour; if one cat in a household growls, the others learn. Another sound not often heard is chatter. When a cat sees something tempting which is out of reach – a bird through a window, for example – it may chatter its teeth together. It is making a killing bite but, because it has

Becoming acquainted with your cat's vocabulary or body language will help you understand its behaviour. This cat is sending out very strong, positive signals. It is happy, confident, and feeling at home in its territory. Note the upright tail, straight back, and relaxed ears. If the cat were feeling out of sorts or worried, its tail would droop and its ears would be held lower.

POWER STRUGGLES

Cats are much more civilized than humans when it comes to fighting. The object of any battle is to determine the top cat, not to inflict unnecessary injury. Cat fights include a great deal of ritual that both protagonists understand, having 'trained' for such battles since they were kittens. The 'undercat' realizes early on that it stands little chance of winning and bows out as soon as possible, the other cat giving it every chance to do so, so that it can fight another day. Neither cat holds any grudge against the other; having worked out where each is in the pecking order, they occupy those places peacefully until the undercat becomes more mature or stronger or the top cat grows older or weaker. Nor does an undercat have to fight every cat in a group; by challenging one or two, it will find its place in the hierarchy and stay there until circumstances alter its placing. In a feral group, the most dominant cat is usually an entire male, but entire females who have had kittens are often high in status and may outrank the males.

Below. Cats do not look very large, frightening or threatening when viewed from the front, so fighting cats will endeavour to show each other a bigger, more menacing and impressive side view. The cat pictured here is exploiting the full range of its aggressive body language. The fur rises along the spine and the tail fur stands out at right angles like a bottle-brush. The eyes narrow into a piercing stare and the feet are poised to take off in any direction at a move from an adversary.

Right. This cat, although smaller, is showing it is no push-over. The ears are back to minimize damage in case of attack, hisses are used as a deterrent and this little cat is ready to roll over and fight while on its back.

Left. Face to face and the tabby, by its upright stance, is already showing its dominance over the ginger cat.

Left. A paw lashes out and the smaller cat ducks. Cats must be allowed sufficient time to sort out a hierarchical structure if they are to live happily with other felines.

Below. These cats are still coming to terms with each other: it may be months before fighting ceases.

TREATING WOUNDS

Cats receive fewer wounds than might be expected through fighting. Submissive cats have wounds on their heels from final swipes as they run away. Dominant cats tend to have wounds on their face from confronting opponents. Wounds that become abscessed should have veterinary attention as antibiotics may be necessary. Simple cuts can be cleaned out with saline solution: use 1 teaspoon of salt to 1 pint of water. Do not use any other antiseptic as many are toxic to cats.

nothing in its mouth, the teeth come together and make a noise.

Purring is not always a sign of contentment. A sick cat purrs when it is in pain and many misguided owners beg vets to save very badly injured cats because they begin purring. An anxious cat may purr with nerves, so when a cat purrs in the veterinary surgery it does not necessarily mean it is happy about the large needle it sees before it.

Sexual Signs

You may not realize immediately when your female cat is in season — but every tom-cat within an 8-km (5-mile) radius will. Her scent will carry across the neighbourhood and the first you may know of her condition is when you see a collection of tom-cats waiting patiently outside your door.

Although it looks as though this tabby-and-white is giving its companion a friendly hug and wash, it is really saying it is boss. By holding its companion down with a paw, it is virtually immobilizing it in the same way as a mother cat would do to her kittens while washing them.

Females can come into season (technically known as oestrus, but also referred to as calling due to the noise made) from as young as four months old. A precocious Siamese may mother kittens while still practically a kitten herself. Longhaired cats may not begin calling until ten or twelve months old while nonpedigree cats usually start the first spring after they reach six months.

Many owners are taken by surprise by the early physical maturity of their female kittens. For this reason, it is wise to take female kittens to the vet for spaying at four to six months, or even earlier for a Siamese. Males need to be neutered at about six to nine months. It is a fallacy that a female needs to have one litter 'for her health', and it is also a myth that males or females feel they are missing something after the operation. The truth is, there are too many cats in the world – rescue shelters everywhere are too full to take in all the unwanted cats – so there is no need to add to their numbers.

If you have not yet made that veterinary appointment, one day you may find your female rolling around as if in pain, or pushing herself along on her stomach with her rump raised in the air, and making blood-curdling, howling noises, particularly at night. She is not in pain, she is simply advertising her availability. Some females are quieter and become more affectionate to their owners when calling, some also urinate or spray around the house. Noisy or quiet, your cat will have only one thing on her mind – to get out of the house and find a mate. She will summon up all her cunning and slip out of a door or window, opened for a split second.

If she is not kept in for the 4 to 10 days of her oestrus, she will conceive, even if she has mated only once – it is rare for cats not to conceive. Female cats (queens) ovulate each time they mate, so if mated with several different tom-cats, they may give birth to a litter of kittens, each having a different father!

Once her first season is over, make an appointment for spaying. Complications are more likely during oestrus, and it is wise to have it done before she comes into season again three weeks later as keeping her indoors is not a good method of permanent birth control. There are also good medical reasons for this, as unspayed cats that do not mate for several seasons tend to develop ovarian cysts. Neutered males also lead a

safer life than full tom-cats. Tom-cats range over wide distances, sometimes getting lost or run over by vehicles in their quest for females. They get into fights and go off their food while other things are on their mind. They also smell terrible. Pheromones – chemical substances in the tom-cat's urine – attract females who find the smelliest males the most alluring.

Males fight not so much for the female herself, but for the temporary right to her territory. She, however, may choose not to mate with the winner, but with another tom, or several others. A very experienced tom tends to mate immediately he gets the opportunity, while a less experienced male may indulge in a courtship of calling and head and body-rubbing before straddling the female. She will raise her rump and the male will grip her neck in his teeth, often breaking the skin (this is one way in which Feline Leukaemia Virus is transmitted). Mating takes just seconds, then the tom leaps out of reach while the female reacts as if in pain and lashes out at the male. The barbs on his penis scrape her vagina as he withdraws, and so trigger the release of a hormone from the pituitary gland that stimulates ovulation.

The female then grooms herself thoroughly and may mate several more times that day and for several days afterwards. After a gestation period of approximately 63 to 65 days, the female produces a litter of, typically, three to five kittens. About a month later, while still nursing the kittens, she comes into season again, ready for mating. Theoretically, a female could have as many as five litters a year.

If your cat is in kitten, the first sign that the birth is imminent may be when she starts looking for a quiet, dark place. You may want to prepare a kittening box in a place that is both comfortable for her and con-

MATING

At a very conservative estimate, at least a quarter of a million unwanted kittens find their way into rescue shelters every year in the UK. Many never leave. In the USA, six million cats are rescued annually by shelters and fewer than a quarter find new homes. The rest are put down. Please do not add to this heart-breaking total by allowing your cat to mate. Have females spayed at four to six months and males neutered at six to nine months. The most common reason given for allowing cats to mate – that it is educational for children – only teaches them to believe that animals exist solely for our benefit.

The photographs used on these pages are stock library shots, which are several years old. No cat was mated or had kittens for the purpose of illustrating this book.

Below. *Feline foreplay can involve rubbing, rolling, purring or even gentle slapping. The female becomes interesting to the male several days before he becomes so to her. He is, therefore, living dangerously in the first few days of his courtship.*

Below. *The female is grasped firmly by the neck while the male manoeuvres into position. The skin may be broken by the male's teeth, which can cause an abscess or transmit diseases such as Feline Leukaemia Virus.*

Right. *As the male withdraws, the female lashes out.*

Left. *After mating, the female always washes thoroughly. Usually the male has retreated to a safe distance by this time, away from the risk of flashing claws.*

Right. *Cats ovulate in response to mating, so the more times the female can mate over the few days she is in season, the more likely she is to have a large litter. Within a short time of mating, she is ready to mate again.*

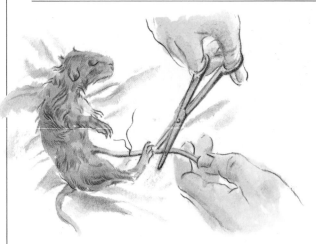

Do not interfere in the birth more than necessary. If, however, your cat has not chewed the umbilical cord leading from the kitten to the placenta within 10 minutes or so, cut it about 3.5cm (1³/₈in) from the navel, having first tied sterile cord around it as shown here.

venient to you, otherwise she may choose the airing cupboard or the bottom of a wardrobe. Cats should not be left alone to kitten, so (unless your cat disappears to give birth) it is wise to keep her company, in case she needs help – particularly if it is her first time. Most cats do have an instinctive knowledge of what to do, but not all, and if there are complications she may need the assistance of a veterinary surgeon.

To make a kittening box, find a large cardboard carton. Cut another one to fit over the first, up-ended to make a 'cave', with an entrance large enough for you to put your hands through, if necessary. Cover the bottom of the box with lots of newspaper for warmth, then several layers of white, paper kitchen towels or pieces of sheet. You can remove one layer at a time, as each gets soiled. Alternatively, you can use pet bedding in artificial fur fabric to give new-born kittens a good surface to grip, but you will need several sheets as it will need washing daily.

Most cats seem to give birth in the early hours of the morning, so try not to sleep too deeply on the night the kittens are due. Your cat may even wake you when the contractions start, but try not to interfere too much – she may just want you to hold her paw. If she seems to be coping well with the birth, leave her to it, but stay within sight if she seems to wish it. In the early stages she may be panting or purring and there may be a clear discharge or a spot of blood. The contractions usually last around 10 to 30 minutes but if they continue for 90 minutes without a kitten being born, call the vet immediately.

When the first kitten is born, the female normally opens its birth sac with her teeth. She may then shear the umbilical cord, or she may wait while she licks her kitten dry. There is no urgency to cut the cord, but if your cat does not open the birth sac, do it for her. Using your fingers, gently pull open the fine, tissue-like membrane, taking care not to nip

This new-born kitten will be licked dry by its mother within a few minutes and will wriggle around to find a teat. New-born kittens start squealing almost immediately after the birth, especially if they cannot find food.

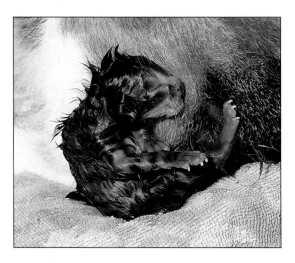

the kitten inside. The mother will lick the kitten's mouth and nose, clearing it of mucus so that it can breathe. If, however, she shows no interest in cleaning the kitten, gently wipe its mouth and nose, and dry its body thoroughly with a clean towel. Ensure that one placenta emerges for each kitten and, if your cat shows

no signs of shearing the cord, do it for her (see illustration far left).

As one kitten is licked dry, the next may appear – the interval between births can be as little as 10 or 15 minutes. If the first kitten has not found its mother's teats by this time, gently place it by a teat to suckle while the mother delivers her second. She may eat the placentas as they appear – it seems they have a function in stimulating milk flow. Allow her to eat one or two – more may cause a stomach upset – and flush the rest away.

The mother's middle teats seem to be favoured and the weaker kittens find themselves on either end. Competition for nourishment is fierce and kittens will trample one another in their quest for food.

New-born kittens knead on either side of their mother's teats to make the milk flow. Even when adult, cats remember this as a pleasurable experience and, every time they are happy, will knead again.

Make sure that the room is warm and draught-free both during and just after the birth – if you are sweating, the temperature is probably about right. New-born kittens lose heat rapidly and need to be cuddled up with their mother. If one of the litter is cold and sluggish, immerse it up to the neck in hand-warm water. If a kitten has difficulty breathing after the birth, hold it at arm's length in the palms of your hands and swing your arms downward, stopping abruptly with the kitten's nose pointing towards the floor. Repeat this several times until the kitten begins to cry and wriggle, a sign that breathing has been restored.

Leave the new family in peace with food and water for your cat and a litter tray within sight of, but some distance from, the kittening box. Some females will not eat or use the tray for a day or two but others are hungry after the birth. Keep visitors away for the first few weeks. If your cat is disturbed and feels her kittens are under threat, she may carry them away and hide them. Keep a close watch on her after the birth and if she seems off-colour or has a pronounced discharge *see your vet immediately*. Some discharge is normal for a day or two, but it is wise to have it checked. Feed her well, let her have as much food as she likes and consider adding a supplement. If you have the slightest doubt take your cat and litter to the vet within 24 hours of the birth for a thorough check-up.

If for any reason your cat cannot or will not feed the kittens, try to find a foster mother – a female who has had kittens around the same time and will accept the new kittens and feed them. The alternative is to feed them kitten-milk formula (from vets or pet stores) from a dropper or an animal-feeding bottle every two hours. You will also have to help them eliminate waste matter for the first three or four weeks, until they can do it for themselves. Rub their tummies gently every couple of hours and wipe their bottoms until they defecate - mother cats perform both tasks with their tongues to keep the nest clean. Finding a foster mother involves less

REMEDIES FOR PREGNANCY, KITTENING AND KITTENS

Read 'Before You Start', pp. 16-17

HOMOEOPATHIC REMEDIES

Caulophyllum (Blue cohosh)
Caulophyllum is one of the best remedies to use during pregnancy. It reduces the risk of abortion and prepares the body for birth by relaxing the tissues. Given to your cat during kittening it will help to stimulate uterine contractions.
SUGGESTED DOSE: During pregnancy: Caulophyllum 30 twice weekly for the last 3 weeks of pregnancy.
During kittening: Caulophyllum 30, every 20 to 30 minutes.

Arnica (Leopard's bane)
Give routinely for 2 or 3 days prior to kittening, during kittening and for a few days afterwards to help prevent bruising, limit bleeding and encourage rapid healing.
SUGGESTED DOSE: Arnica 30, twice daily.

Urtica urens (Stinging nettle)
Urtica in high potency can stimulate milk flow in cases where the queen is not producing enough to feed her kittens. In low potency it has the opposite effect and can be used to dry up milk prior to spaying.
SUGGESTED DOSE: Urtica 200, twice daily for 2 days to encourage milk flow. Urtica 3x, twice daily for 5 days to dry up milk.

Sepia (Ink of the cuttlefish)
Sepia can help in situations where the queen rejects the kittens or will not allow them to feed. Platina (the metal) is a good alternative especially suited to Siamese cats.
SUGGESTED DOSE: Sepia 30, or Platina 30, twice daily.

Kali carb (Potassium carbonate)
General debility following abortion, kittening or nursing may respond.
SUGGESTED DOSE: Kali carb 30, twice daily.

The following are all valuable first-aid remedies for treating new-born kittens. Use the 30th potency given every 10 minutes.

Arnica montana (Leopard's bane)
This is a useful remedy for treating cases of injury and shock.

Hypericum (St John's wort)
Indicated where limbs have been crushed.

Helleborus (Snow rose)
This is useful for weak, depressed and sluggish kittens that are slow to come to after delivery.

Antimony tartrate (Tartar emetic)
Use where there is rattling in the chest due to mucous inhalation.

Carbo veg (Vegetable carbon)
This remedy may be a life saver when given to cold and collapsed kittens.

HERBAL REMEDIES

INTERNAL
Raspberry leaf (Rubus idaeus)
This is a traditional remedy used during pregnancy to tone and strengthen the uterus. It also encourages strong contractions and helps limit any haemorrhage. Use a proprietary preparation on a daily basis throughout pregnancy.

Squaw vine (Mitchella repens)
An American Indian remedy which helps prepare the uterus for kittening. It can be given during the last 4 or 5 weeks of pregnancy together with Raspberry leaf.

BACH FLOWER REMEDY
Rescue Remedy
This can be given to mother and kittens to help deal with the shock of the delivery.

time and effort and will feel more natural to the litter. If you want a kitten to keep your cat company, do not encourage her to have a litter – it is much less bother to buy a kitten.

Problem Solving

The most common problem cat owners face is inappropriate spraying, urinating and defecating around the home. People consider such behaviour 'dirty' but to a cat it is natural – they are simply using urine and faeces as territorial markers. When a cat sprays in the home it is signalling its ownership in exactly the same way that we do when we write our names in the flyleaf of a book. It is worse than pointless to chastise or smack such a cat – it will only make it nervous and insecure and more likely to repeat the behaviour. Aggression, too, is becoming a more common problem as cats are expected to live closer together on ever-shrinking territories. It may be directed towards other cats, or towards owners, or both. It is important to find the root cause of the problem, which may involve working through all the possibilities first.

Indiscriminate Urination or Defecation

Illness Very often when a cat soils in the home, it is the first sign of illness. As some disorders can rapidly become life-threatening, it is vital to call the vet the same day. If a cat has cystitis, a lower urinary tract disease (FUS) or kidney problems such as nephritis or impacted anal glands, soiling may be the only symptom until the condition becomes very serious indeed. So regard it, first of all, as a cry for help.

Stress If your cat is under stress from a change in routine, too many cats in the household or from illness, it may soil. If possible, change whatever is causing the stress. If

this is not possible, be patient and try to make your cat feel secure – the behaviour should cease when it feels less threatened.

Tummy upset If your cat has eaten something that disagrees with it, it may simply not be able to get to its tray in time. If it has diarrhoea, ensure that it has plenty of water to drink (no milk) and do not feed it for 24 hours. Overfeeding can cause diarrhoea.

Old age Many cats can begin to lose control of their bodily functions as they age. They may not have enough time to get outside or to the litter tray and often become very upset by these 'accidents'. Provide at least two litter trays – one upstairs and one downstairs if you live in a house, or one at each end of an apartment. Ensure that your cat has regular veterinary check-ups – at least once a year but preferably twice – as its other bodily systems may not be functioning quite so well either.

Lack of litter tray Cats are sensible creatures. They do not like going outside in sub-zero temperatures, in wind or rain, or squatting on ice or mud. And the older they get, the more they will refuse to do so. Buy a tray.

Dirty litter tray Cats do not like to use an already-used tray. Ideally, remove the faeces as they are deposited, or else twice a day.

Repellent litter tray Trays need to be washed regularly, but if they are washed or disinfected with something that repels the cat (such as pine-scented cleansers) it will not use them. (See p. 34 on cleaning the litter.)

Wrong type of litter Cats have their own preferences so if they object stop using litter deodorant or deodorized litter and try a different type (see p. 32).

Lack of privacy Place the tray in a quieter area or buy a covered tray.

Confusion Some cats 'cover up' the contents of their litter tray with their front paws from outside the tray. Because they feel carpet under their paws, they may use the carpet next time. Place the tray inside a cut-down cardboard carton or on a sheet of plastic. Cats do not usually soil on plastic as they do not like the feel of it, so leave sheets wherever the cat is soiling regularly.

Other causes Females in season (and males responding to them) may soil as well as spray around the home, to mark their territory. Nervousness is another cause, and the only answer to this is patience. Foods, too, can cause soiling in susceptible cats – if you suspect this, change the diet to see if the behaviour stops. Other substances in the cat's environment – new paint or a new carpet-cleaning treatment for example – may provoke soiling as an allergic reaction.

Whatever the cause, the soiled area must be thoroughly cleaned – if any trace of the smell remains, the cat will use the area again. Wash the area with clean water, then, after testing for colour fastness, wash with a weak solution of environmentally friendly bleach and water. Alternatively, spray with one of the deodorizing sprays available from vets or pet stores, preferably one with a bacterial action that 'digests' odours, or a natural mineral deodorizer such as zeolite which can be sprinkled over the source of odour and vacuumed up later.

If an area is soiled frequently, try a repellant such as a vinegar rinse, oil of peppermint or citrus oils, but test for colour fastness if you are applying it directly on to a carpet or furnishings. Alternatively, you could cover the area with plastic sheeting, a chair or even

your cat's feeding bowl, as it will not soil where it eats.

One persistent soiler refused to use anything other than carpet. In desperation, its owner placed a square of carpet in its litter tray and the problem was solved. Each time she replaced the carpet, she surrounded it with more cat litter, until there was no carpet left and her cat was using litter again.

Spraying

Territory Cats spray to stake their claim to territory. Females and neutered cats who have smaller territories are less likely to spray than tom-cats. In the male, the levels of the hormone testosterone, which can trigger spraying behaviour, are lowered when the cat is neutered, but if some testicular tissue is left after the operation, the cat will continue to spray. This, however, is very rare.

Sexual advertisement Spraying is designed to attract the opposite sex. A male may spray in response to a female living some distance away as he will be able to smell her when she is in season, and he may even spray *over* a receptive female. Unless the cat is required for breeding, neutering is the answer.

Overcrowding If a cat feels threatened on its own turf, either by an aggressive cat or by too many other cats it may resort to spraying in an attempt to assert itself. Try to give your cat more space – perhaps somewhere to sleep on its own. Frequent spraying tends to occur when there are four or more cats.

Stress Stress of any kind – alteration in routine or illness – can cause spraying. Again, try to alleviate the cause of the stress. Stress is a greater problem for cats than most people realize. As creatures of habit, they hate any alteration in their routine, even

REMEDIES FOR BEHAVIOURAL AND PSYCHOLOGICAL PROBLEMS

Read 'Before You Start', pp. 16-17, and see also, Remedies for Stressful Situations, p. 55.

HOMOEOPATHIC REMEDIES

Unless otherwise indicated, use the 30th potency, twice daily, changing to the 200th, 3 times weekly, where a response is observed.

Aconite (Monkshood)
The principle indications for aconite are fear, anxiety and restlessness. The cat is easily startled and may avoid human contact. Effective after a fright.
SUGGESTED DOSE: Aconite 1M, 2 doses 1 hour apart.

Calc Carb (Calcium carbonate)
A remedy for slow, sluggish kittens that are late developers. It is also worth trying in cases of depraved appetite where there is a liking for wool and other indigestible articles. Calc Phos,

Cina, Nitric acid and Alumina can also be useful in treating this problem.

Hyoscyamus (Henbane)
One of the main remedies for treating aggression, particularly where a cat will seek others to attack and is extremely quarrelsome. Jealousy, suspicion and excitability are also notable features.

Ignatia (St Ignatius bean)
The main use of this remedy is in treating the ill-effects of grief, for example after the loss of a companion animal, or pining. Where Ignatia fails; Natrum Muriaticum may work. Other indications for Ignatia include hysteria and depraved appetite.

SUGGESTED DOSE: For grief and pining: Ignatia 1M, 3 doses given over a 24-hour period, or Nat mur 30, twice daily for 10 days.

Staphysagria (Stavesacre)
It is most valuable in treating conditions which arise after an injury or an operation where an aura of resentment (the keynote of Staphysagria) exists. Useful post-operatively where there is pain and in treating pent-up anger. Spraying in males which occurs after neutering often responds.

Ustilago Maydis (Corn smut)
Another remedy for spraying where territorial marking is the root of the problem.

HERBAL REMEDIES

INTERNAL
Valerian (Valeriana officinalis)
Valerian is well known for its sedative and calming qualities. It is non-addictive and safe to use in treating general nervousness, hysteria, excitability, restlessness and some forms of epilepsy.

Skullcap (Scutellaria laterifolia)
A widely used herbal sedative, usually given in combination with Valerian. Useful in treating hysteria and epilepsy in its own right.

Hops (Humulus lupulus)
Hops also have a sedative action, calming nervous animals by having a general relaxing effect on the nervous system. They combine well with both Valerian and Skullcap.

EXTERNAL
Catnip (Catmint, *Nepeta cataria*)
Whilst acting as a mild sedative for humans, this remedy has the opposite effect on cats. Bags of bruised leaves can provide an uplift for depressed cats, encouraging them to play.

BACH FLOWER REMEDIES
Rescue Remedy
Use in any acute situation where there is shock, fright, fear, panic or pain.
Dose every 15 minutes.

Mimulus (Mimulus guttatus)
Indicated for fear of known things such as thunder or a visit to the vet. Suits timid, shy cats. Dose 4 drops 4 times daily.

Holly (Ilex aquifolium)
Associated with jealousy and suspicion. Use when introducing a new cat into the household and to treat aggression.

something as minor as having their feeding times changed. Exercise, massage and play (see Chapter Two) will help to relieve the stress particularly if there is catnip in the toys or sprinkled on the floor. An excellent diet (see Chapter One) will also help.

Aggression

Illness or injury If a cat feels ill, or if its owner inadvertently touches an injury, it may instinctively lash out. A trip to the vet to treat the problem will prevent a repetition.

Previous ill-treatment A cat that has suffered ill-treatment is more likely to be nervous or indifferent but some do become aggressive. Understanding and patience are the only solutions while the cat comes round in its own time. Try not to force the pace. Let the cat decide when it will be friendly – it may not take long, but it *could* take several years!

Instinctive reaction Many cats bite if their bellies are stroked. The movement, similar to the scrabbling of claws on the belly by a fighting cat, triggers a fighting reaction. Some cats bite if they are stroked anywhere for too long. Stroking is a superior gesture – mother cats lick their kittens, the earliest form of 'stroking' a cat experiences – and a cat may become annoyed if it is forced to retain this submissive position for too long. Watch for the twitching tail and the ears going back, then stop. Instead of biting, some cats may hold a wrist or hand in their teeth, to restrain it, as they would restrain a kitten, by gripping it around the neck. As with holding a kitten, the intention is not to hurt it or break the skin.

Poor introduction to a new cat or kitten When introducing your cat to a new feline companion, do not cause jealousy and resentment by making a fuss of the new-comer. The golden rule is: *always let your cat make friends with its new companion before it sees you doing so.* This is also true of introducing a dog or puppy, which should be kept on a lead at first.

Other reasons If two cats or a cat and dog get along well, but one is away from home for a time, fighting may break out on its return. They usually settle down together again, given time. Toxins in a cat's environment or additives in its food may also provoke an aggressive reaction. Try changing your cat's diet and use as few chemicals as possible around the home. Stress, too, of any sort, may also cause aggression (see p. 55 for ways of combatting stress). If other methods of coping with aggression fail, your vet may prescribe a mild tranquillizer.

The more you understand your cat by listening and watching its unspoken language, the more you will be able to pre-empt problems, and the faster you will respond to any that do occur.

CHAPTER FOUR

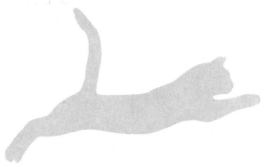

General Care

Prevention is always better than cure and a little preventative health care will help keep your cat happy and in good health. Another benefit is that it will save you money as your cat will not need as many trips to the veterinary surgery.

Grooming your cat daily is a quick and easy way to spot the presence of fleas or other parasites before they have time to take hold and become a major problem. By taking immediate action (see pp. 89-91), you will minimize the risk of parasite-induced illnesses and allergies that are not only very uncomfortable and possibly dangerous for your cat but may require regular and ongoing veterinary attention. Not least of its benefits, grooming also gives you an excellent opportunity to spend some quiet time with your cat in an occupation that you will both, hopefully, enjoy.

The amount of time needed to clean ears and eyes will vary from cat to cat and from

Left. A young cat like this one, with short fur, is able to groom itself, although you should help with a weekly combing to prevent hairballs, Longhaired, old or sick cats must have regular assistance with grooming.

breed to breed. Some breeds, such as the Persian, sometimes suffer from blocked tear ducts, and may need their eyes wiped daily, while a mixed-breed cat may never need attention to its eyes throughout its lifetime. However, all cats who will permit it will benefit from having their teeth cleaned; it prevents the build-up of tartar that can lead to stained teeth, painful gums, difficulty in eating and possible extractions in later life. Claw-trimming may be more for your benefit than the cat's, but it can help to keep relations positive between you. Whether or not you choose to follow the recommendations that are given in this chapter will depend on your cat's individual needs and preferences, and your own.

Claw Care

Cats' claws are constantly growing. The old, worn-out sheath is removed by stropping or ripping it off with their teeth, leaving a sharp, new claw underneath. When owners are concerned that their cat has a lost claw, it is likely that they are, in fact, mistaken: what they have found is simply the claw sheath.

Scratches are painful and can damage the

No matter how much you trim your cat's claws to blunt them, it will sharpen them up again in a few days by stropping on a scratching post or tree. Claw-trimming is not traumatic for the cat, although the owner may sometimes find the experience more difficult.

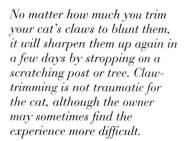

furniture. They can be avoided, however, with no trauma to the cat, by simply trimming the claws using cat-claw clippers or human nail clippers or scissors (dog clippers are too large). Hold your cat firmly (if it objects, enlist the help of a friend to hold while you trim), place your thumb on top of the paw and your fingers underneath. Press gently and the claw will unsheath. Take off just the tip of the claw – about 3.5 mm (⅛ in) – and be careful not to cut down to the quick, which will cause copious bleeding. If you are uncertain, ask your vet to demonstrate.

Claw-trimming is not the same as de-claw-

Middle right. Press your cat's paw between your finger and thumb to unsheathe the claws. Bottom right. Trim just enough to blunt them, about 3.5mm (⅛in) from the tip. Only trim the front claws. If your cat objects to claw-trimming, wrap it firmly in a towel, leaving only the head and one paw exposed. Hold it while a helper trims the claws.

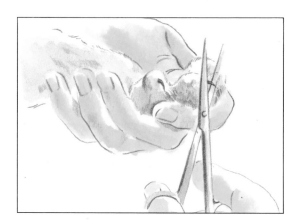

ing. This involves removing the entire claw and the first joint down to the knuckle and can cause post-operative trauma, infection and behavioural problems. The procedure is discouraged by the British Veterinary Association, The American Humane Association and the Canadian Veterinary Medical Association. As an alternative to de-clawing, vets provide vinyl claw caps or beads that fit over the claw.

Eye Care

The cat has a third eyelid, the nictitating membrane, which is sometimes seen at the corner of the eye. It can be a first sign of illness or eye injury. However, it may just mean that the cat has lost weight: the eye rests on a small pad of fat which can shrink allowing the third eyelid to pass across the eye.

Above and below. *The cat's pupil contracts in bright light to filter out some of the light (see above) and expands in the dark (see below) to let in more light and help the cat to see.*

Above. *Most cats manage to keep their own eyes clean, but sometimes wiping is needed. Dissolve a teaspoon of salt in a pint of lukewarm water, wet a pad of cotton wool with the solution and wipe gently over the eye. If there is a cloudy discharge or the cat keeps an eye closed or paws at it, do not bathe it but consult a vet immediately (see Care of the Eyes, p.78).*

CARE OF THE EYES

Read 'Before You Start', pp. 16-17

HOMOEOPATHIC REMEDIES

Arnica montana (Leopard's bane)
First-aid treatment for any eye injury. Prompt use will reduce pain, inflammation, bruising and prevent haemorrhage.
SUGGESTED DOSE: Arnica 30, 2 or 3, doses 1 hour apart.

Symphytum (Comfrey, Knitbone)
One of the best remedies to use in treating blows to the eye from blunt objects. It has a significant effect in reducing pain.
SUGGESTED DOSE: Symphytum 30, 2 or 3, doses 4 hours apart.

Ledum palustre (Marsh tea)
Use where there has been a puncture wound to the eye from a sharp object, such as a thorn or a claw. Ledum is also of value where the eye has been severely bruised and where haemorrhage has occurred.
SUGGESTED DOSE: Ledum 30, 2 or 3, doses 4 hours apart.

Apis mel (Honey bee)
Apis is useful in treating acute eye infections, where the conjunctival membranes become puffy, very inflamed and swell, almost obscuring the eyeball. There is a profuse discharge and strong light is resented.
SUGGESTED DOSE: Apis 30, 3 times daily.

Pulsatilla (Wind flower)
Use in cases of conjunctivitis where there is a profuse sticky, creamy, yellow discharge which glues the lids together.
Particularly valuable in treating kittens with gummy eyes.
SUGGESTED DOSE: Pulsatilla 30, twice daily.

Merc sol (Mercury, Quicksilver)
Chronic conjunctivitis often responds to this remedy. The discharge from the eyes is greenish and worse overnight. Eyelids are red and swollen and bright light causes discomfort. Very severe symptoms, particularly where ulceration is present, indicate Merc corr (corrosive sublimate of mercury).
SUGGESTED DOSE: Merc sol 6 or Merc corr 6, twice daily for 10 days.

HERBAL REMEDIES

INTERNAL
Cleavers (*Galium aparine*, Goosegrass)
Can be used to help clear chronic conjunctivitis. It does this by toning the lymphatic system, strengthening the liver and cleansing the blood.

Eyebright (*Euphrasia officinalis*)
As its name suggests Eyebright can be useful in treating eye conditions. Valuable in both acute and chronic eye infections, it has a powerful anti-catarrhal effect, cools the blood and detoxifies the liver.

Golden seal (*Hydrastis canadensis*)
Helps in clearing sticky eye discharges and in chronic eye conditions. It can be combined with Eyebright.

EXTERNAL
Eyebright (*Euphrasia officinalis*)
Externally Eyebright can be used to treat almost any eye problem, including conjunctivitis and keratitis. Steep a half-teaspoon of the dried herb in a cupful of water for 10 minutes or dilute 1 or 2 drops of mother tincture in an eggcupful of cold boiled water.

Calendula (Marigold)
Diluted Calendula lotion can be used to bathe the eyes where there is a mild infection. In combination with Hypericum it reduces pain after eye injury.

The function of the eyelids is rather like that of a car's windscreen wipers; they sweep across the eye, brushing aside debris and keeping the cornea moist. Cats have a third eyelid at the inner corner of each eye that performs the same function and is not usually visible. If it becomes visible, veterinary attention should be sought.

If it is necessary to clean your cat's ears, do so very carefully without pushing the cotton bud down into the ear canal. If the cat is scratching at its ears or shaking its head, veterinary attention should be sought (see Care of the Ears, p.80).

Healthy eyes need nothing more than an occasional wipe with damp cotton wool. If you notice anything unusual in the eye area - cloudiness, discharge, squinting or the third eyelid, *see your vet immediately.*

If an eye becomes dislocated from its socket, do not attempt to replace it. Keep the eye wet with lukewarm water and see your vet as an emergency patient.

Ear Care

Some cats have ears that never need to be cleaned; others, notably some pedigree breeds, need their ears cleaned at least every few days.

To clean away wax use a cotton bud, taking care not to push it into the ear canal. Or use cotton wool, moistened with vegetable oil, which will not hurt your cat but is unlikely to clean the ears as well as a cotton bud.

CARE OF THE EARS

Read 'Before You Start', pp. 16-17

HOMOEOPATHIC REMEDIES

Aconite (Monkshood)
Use for acute inflammation where the ear flap or canal is swollen and hot. Aconite needs to be given early on in order to be effective.
SUGGESTED DOSE: Aconite 30, every 2 hours.

Belladonna (Deadly nightshade)
Useful in acute ear infections where much pain is apparent and the ear is very red and inflamed. The patient is angry and irritable. If the symptoms match, Belladonna can be used to treat bite wounds to the ear.
SUGGESTED DOSE: Belladonna 30, 3 times daily.

Hepar sulph (Calcium sulphide)
More established ear infections may respond to this remedy. There is a suppurating discharge full of pus and the cat resents the ear being touched.
SUGGESTED DOSE: Hepar sulph 30, twice daily.

Graphites (Black lead)
The guiding symptom to the use of Graphites is the presence of a sticky glutinous honey-like secretion in the ear canal or on the ear flap. It tends to suit cats of a timid nature.
SUGGESTED DOSE: Graphites 30, twice daily for 1 week.

Sulphur (The element)
A good remedy to use if ear mites are present. The ear smells and is red and inflamed, causing the cat to scratch. The skin generally looks unhealthy but the most important guide to using Sulphur is the patient's avoidance of heat.

SUGGESTED DOSE: Sulphur 30, twice daily for 10 days.

Psorinum (Scabies vesicle)
In contrast to Sulphur, the Psorinum patient seeks heat. Irritation from the ears is intense and there is usually a brown, offensive smelling discharge. Additionally the coat will have an untidy appearance.
SUGGESTED DOSE: Psorinum 30, once daily for 5 days.

Tellurium (The metal)
This is an effective treatment for eczema of the ear flap, where the edges are scabby and the surface is scaly. Lesions are often circular. Where there is a discharge it is offensive and worse on the left side.
SUGGESTED DOSE: Tellurium 30, twice a day for 5 days.

HERBAL REMEDIES

EXTERNAL
Calendula (Marigold)
Diluted Calendula lotion is ideal for cleaning out sore and infected ears. As well as being anti-inflammatory and anti-fungal, it is soothing and encourages healing. Witch hazel can be used in a similar way.

White horehound (*Marrubium vulgare*)
An infusion of a half teaspoon of dried Horehound in a cupful of boiling water and allowed to cool also makes a useful ear cleaner in cases of infection.

Aloe (*Aloe vera*)
In white cats the non-pigmented ear tips are prone to sunburn and blistering, leading to possible cancerous changes. Aloe gel soothes inflamed ear flaps and reduces the discomfort. St John's wort oil can be used as an alternative.

Almond oil
This remedy can be used to clean out waxy ears. It softens and loosens wax, and soothes inflamed skin. Olive oil can be used as an alternative but is not quite so effective.

ESSENTIAL OILS
There are several essential oils that can be used effectively in the treatment of ear problems. In particular Eucalyptus, Lemon and Rosemary are insect repellant and useful in ear mite infestations. Tea tree oil is a powerful antiseptic and Thyme has a similar but less potent effect in treatment of infections.
SUGGESTED DOSE: Use 1 or 2 drops of an appropriate oil mixed in 10 ml of almond oil. Instil a few drops of this into the ear once daily and massage well. Treat for about 10 days.

When a cat yawns, it is usually interpreted as a reassuring sign by other cats. It means that its intentions are friendly. A feline yawn can also be used to show tiredness, as this cat is doing.

If your cat persistently shakes its head and scratches, it may have ear mites. Do not treat them yourself but see your vet. Likewise, consult your vet if your cat's ears smell, are reddened inside, have a brown, waxy discharge, or if the ears have been sunburned (especially on a white cat).

Mouth Care

Cats, like humans, only ever have two sets of teeth, so their care is vitally important. Although some vets are now offering feline fillings and caps for cat teeth, dentistry for our pets is not new. According to the rumours, at least one 19th-century cat had a set of false teeth made of ivory. For the majority of cat-owners, however, the most effective and inexpensive tooth care is still simple brushing to keep tartar at bay, although this never occurs to many owners.

A kitten's 'milk' teeth erupt through its gums during the first four weeks of life and by six to eight weeks it will have its full complement of 25 teeth. Teething problems may occur at around three to seven months when the permanent teeth come through, pushing out the first teeth. These may drop out but are more likely to be swallowed harmlessly.

Adult cats have 30 teeth: 4 large canines which hold and kill prey, 12 incisors between the canines which are used for gnawing, and 14 back teeth (pre-molars and molars) which act like scissors, shearing meat into pieces.

These teeth are designed to work hard, so cats fed entirely on soft, processed food may suffer from a build-up of tartar. This can be minimized by feeding cats a little dry food daily and by giving them an occasional chunk of raw meat.

Before attempting to clean your cat's teeth, allow your cat to get used to your fingers in its mouth. Once a day for several weeks, simply raise the lips by placing your finger and thumb on either side of the mouth and lifting. When your cat no longer objects to this, run your finger along its teeth. Only then is it time to use a toothbrush – use a child's or buy a pet toothbrush from your vet. Then add toothpaste – preferably a meat-flavoured variety available from vets – or baking soda, salt and water or the juice from a crushed clove of garlic added to a cupful of water. Do not use toothpastes formulated for human use.

Tartar on the teeth and red, sore gums. Gingivitis can cause bad breath, dribbling and make a cat lose weight as eating will be too painful. Prevention, as always, is better than cure, as treatment may involve an expensive operation.

Grooming and Coat Care

There are up to 18,000 hairs per cm^2 (120,000 hairs per in^2) on a cat's belly and around 9,000 per cm^2 (60,000 per in^2) on its back. They divide into three types – a thick topcoat which is relatively weather-proof, a downy undercoat to provide insulation and guard hairs which are sparse but coarse.

Cats' coats are thickest in winter when muscles at the hair roots contract to fluff out the fur, trapping a layer of air which keeps the cat warm. In summer, moulting is triggered by the longer days of spring (not by the warmer weather) and the hair thins. There is always a small natural hair loss but if cats are under stress, ill or old they will moult at any time of year.

Healthy cats have glossy fur which, when parted, shows clean, smooth skin. A 'staring'

Feline teeth-cleaning is much simpler when the correct equipment is used. Animal toothbrushes and meat-flavoured toothpastes are available from veterinary surgeries. First of all, accustom your cat to having its mouth touched by your bare, clean fingers. Only when it does not object to this, try gently brushing with a dry brush. Then add toothpaste or an appropriate cleansing solution.

CARE OF THE MOUTH

Read 'Before You Start', pp. 16-17.

HOMOEOPATHIC REMEDIES

Fragaria (Wood strawberry)
A useful remedy both for the prevention of tartar and in removal of deposits already present.
SUGGESTED DOSE: Fragaria 30, once weekly.

Merc sol (Mercury, quicksilver)
Valuable in the prevention and treatment of gingivitis, particularly where associated with kidney disease. Symptoms include salivation, spongy bleeding gums and bad breath.
SUGGESTED DOSE: Merc sol 6, twice daily for 14 days.

Nitricum acidum (Nitric acid)
Suitable where ulcers are present either on the soft palate or tongue. Also in gingivitis where the teeth are loose and the gums are soft and bleed. An excellent remedy in treating rodent ulcer which often occurs on the lip margins. Local application of Galium aparine (Cleavers) tincture can also help this problem.
SUGGESTED DOSE: Nitric acid 30, twice daily for up to 3 weeks.

Phosphorus (The element)
Indicated where the gums bleed persistently, are swollen and ulcerated. Phosphorus given after tooth extraction can halt bleeding. Arnica given routinely before and after dental work can prevent this occurring.
SUGGESTED DOSE: Phosphorus 30, twice daily for 10 days.

Calc fluor (Calcium fluoride)
Indicated where the teeth are loose and are of poor quality. Calc fluor may prevent further deterioration and improve the strength of the enamel.
SUGGESTED DOSE: Calc fluor 30, once weekly.

Calc carb (Calcium carbonate)
This is a valuable remedy where there is impaired calcium metabolism, which could lead to the development of poor teeth. Kittens needing this remedy tend to be overweight, lazy and slow to develop. They sometimes have an unusual craving for indigestible objects.
SUGGESTED DOSE: Calc carb 30, once weekly for 6 weeks.
If the kitten is lean and bony, an equivalent dose of Calc phos (Calcium phosphate) is recommended instead.

HERBAL REMEDIES

INTERNAL
Garlic (Allium sativum)
Daily doses of Garlic can help prevent mouth infection and control gingivitis.

Nettles (Urtica dioica)
Nettles have a potent diuretic effect and help cleanse the body of impurities. They can be of great benefit in cats with chronic gum disease arising from kidney problems. Add a pinch of dried Nettle powder to each meal.

Echinacea (Echinacea angustifolia)
Echinacea is an effective anti-bacterial remedy in treating gingivitis and other infections of the mouth. The tincture can be given internally to your cat whilst a decoction can be used to swab the mouth, though this method may cause salivation in some cats.

EXTERNAL
Golden seal (Hydrastis canadensis)
Add one half teaspoon of the powdered herb to a cup of boiling water and allow to cool. Daily cleaning with the infusion will reduce infection in sore and inflamed gums.

Myrrh (Commiphora molmol)
This is an effective remedy, which has both anti-bacterial and astringent properties. It can be used to treat a variety of mouth conditions including gingivitis. One half teaspoon of the powdered resin should be added to a cup of boiling water. Infuse for 15 minutes, strain and cool before bathing your cat's gums.

Calendula (Marigold)
Diluted Calendula lotion can be used as a mouthwash to bathe sore gums and to assist healing after dental work.

Left. *This cat is displaying the only grooming aid it possesses – its tongue. Rough barbs on the tongue act like the teeth of a comb, parting and smoothing the fur. But if any hair is left on the tongue, the cat cannot spit it out and must swallow it. Swallowed hair can form a ball and an internal blockage, so regular grooming and hairball treatments when necessary will help keep a cat healthy.*

Below. *Do not assume that a cat has no need for occasional hairball treatments just because it is shorthaired. Even the cats with the shortest, sleekest fur can have problems with hairballs. These can be the result of regular grooming of other cats in the household – as demonstrated here – especially if those cats are longhaired.*

coat, when the fur is rough and unkempt-looking with no shine, is a sign that all is not well – it may indicate a poor diet, worms, illness or simply old age. Cats groom their fur with their tongues, the barbs acting as very efficient combs. Those parts of the body the cat cannot reach, such as the face, it wipes over with the side of a licked front paw.

Some fur is swallowed every time a cat grooms, forming hairballs in the throat or gut. These may be coughed up or passed in faeces but if not, they will collect in the bowel where they may form a blockage, requiring surgical removal. To prevent possible problems, groom your cat regularly and, to ease any swallowed hair through the body, feed it an oily fish such as sardines or mackerel once a week. There are also proprietary hairball remedies, many based on malt, which taste attractive to cats. Give them when necessary, but not more often than recommended on the container. Another remedy is medicinal liquid paraffin dribbled gently into the side of the mouth. However, as it coats the cat's stomach and hastens the passage of food, robbing the cat of nourishment, only admini-

ster when necessary – about one teaspoon per day for three days no more often than once a month. If a suspected hairball remains unexpelled, consult your vet.

Hairballs can be minimized by giving a cat some help with grooming. Even shorthaired cats benefit from thorough combing once or twice a week, especially if they are elderly or if they live with longhaired cats. Longhaired cats need combing more frequently, ideally a couple of times a week. Persians should be groomed every day – some owners spend as much as 20 minutes a day grooming Persians. If cats groom each other they can end up with twice the amount of swallowed fur, so if you have more than one cat, make a special point of grooming.

Use a metal comb with equal-length teeth

Below left. *Healthy fur should be glossy, smooth and lie flat when stroked or combed, such as the coat shown here.*
Below. *The fur shown here belongs to a cat that is moulting. Although this is a perfectly natural process, the fur in this picture closely resembles that of an unhealthy cat. It is a 'staring' coat that has no shine and feels rough to the touch.*

GROOMING

Grooming should not be left entirely to your cat. The only tool a cat has for grooming is its tongue, which, although effective in combing, has a built-in disposal problem. Any fur that catches on the tongue's barbs cannot be spat out and must be swallowed. This can build up and form a furball in the bowel if the cat cannot eliminate it.

By combing your cat regularly, preferably daily, you are not only improving its appearance, but also its health and well-being. Kittens are first groomed by their mother when they are too young to groom themselves. So if you take over that role, you are providing a comforting and calming routine that reminds your cat of its carefree early days. You are also, tactfully, pointing out that you are the boss.

Below. All cats, even the shortest-haired, should be groomed regularly by their owners and this cat is showing how much it is enjoying the experience by kneading or paddling. A steel comb is the only equipment really necessary, although a brush has a soothing feel equivalent to a massage; either one is more efficient than a cat's tongue as loose hairs can be removed. Use a pure bristle brush to avoid 'fly-away' fur.

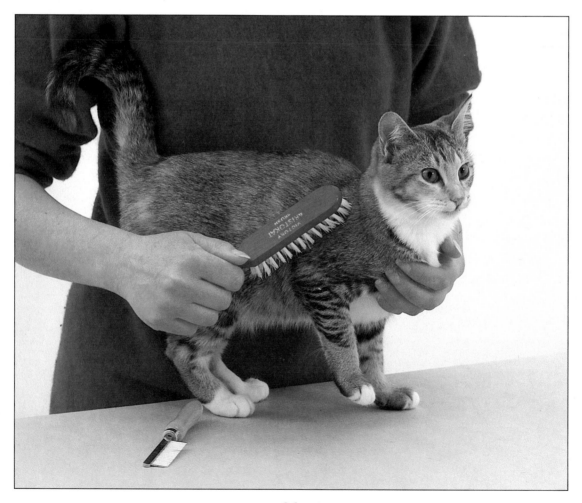

FURBALLS

With furballs, also called hairballs, prevention is always better than cure. Regular, preferably daily, combing will prevent a build up of swallowed hair and save a fortune in vet's fees. Feed your cat oily fish at least once a week and ensure it has access to grass. You may not be charmed when your cat expels a furball but it saves money.

Below. *Cats will lick all the parts they can reach. Paws will be licked and used to wipe over difficult-to-reach areas, such as the head. Fifty years ago, experts recommended feeding cats butter to act as soap in their ablutions!*

Above and below. *Use a metal comb with alternate long and short teeth for a longhaired cat. Comb small sections at a time to prevent tugging and hurting your cat. Matted hair should be cut out carefully or 'unpicked' gently using an implement available from sewing shops designed to unpick embroidery stitches.*

for a shorthaired cat and a wide-toothed comb with alternate long and short teeth for a longhaired cat. Before buying a comb, run it through your own hair to ensure the teeth are not too sharp. This is all the fur-grooming equipment you really need but a natural bristle brush can be used to finish off.

Grooming is an extension of stroking and should be a pleasure. Start when your cat is a kitten by simply stroking over its fur with bare hands, choosing times when your kitten is relaxed but not asleep. After a few days, you can very gently begin to stroke with the comb. If your kitten shows any tendency to play with the comb, or if it objects to the combing, stop immediately and try again another day. When grooming, start with the head, back and tail, then the chest. If your kitten is lying on your lap, you may be able to turn it over to comb its belly. If not, stand the kitten on a table and comb the belly from each side in turn. Use short strokes and, if the comb catches in fur, ease it out gently without tugging. If there is a knot of fur, try to unpick it with your fingers. If this is impossible, cut it out using round-ended scissors and cutting away from your cat's skin towards the end of the fur. Some long fur tangles more easily than others and combing may be easier if you sprinkle a little talcum powder, fine fuller's earth (cat litter) or a proprietary grooming preparation on it. Start and finish grooming with an area your cat particularly enjoys having stroked, say, under the chin or behind the ears, so that your cat remembers each session as a pleasureable experience.

Cats rarely need bathing but a grubby cat can be given a dry shampoo by sprinkling talcum powder or fuller's earth on the coat and brushing out thoroughly. Alternatively, bran warmed in an oven can be rubbed into the coat and combed out, but ensure it is not so hot that it burns your cat.

At the first sign of scratching, comb your cat and inspect the contents of the comb for signs of fleas. Treat immediately if they are present before an infestation starts to get out of control or your cat develops an allergy.

If this is not enough and your cat needs a bath, use a proprietary cat shampoo or a baby shampoo but avoid anti-dandruff shampoo or detergent. Gather together the shampoo with several dry towels and unbreakable jugs in a warm bathroom or kitchen. Place several inches of hand-warm water in the bottom of the sink and pour in a little shampoo to break the surface tension, so that you can wet your cat down to the skin, and have hand-warm rinsing water nearby. If your cat is likely to object to its bath, trim its claws and, if necessary, have someone to help you with the bath. You can place some

REMEDIES FOR SKIN PROBLEMS AND FLEAS

Read 'Before You Start', pp. 16-17.

HOMOEOPATHIC REMEDIES

Sulphur (The element)
The cat that needs Sulphur has a poor coat and is lazy. Fleas are usually in evidence and the skin feels dirty, lacks condition and looks unhealthy. The ears and anal region are often inflamed and itching is prominent. Symptoms are worse for warmth and in the evening. The most important guide to the use of this remedy is the avoidance of heat. Sulphur can also be given on a regular basis to deter fleas.
SUGGESTED DOSE: Sulphur 30, twice daily for 10 days.
To deter fleas: Sulphur 200, twice weekly for 1 month.

Natrum muriaticum (Sodium chloride, common salt)
Greasy looking hair with either oily or dry skin suggests Nat mur. Crusty eruptions in the bends of the limbs and itching leading to tugging out of hair can be features. The remedy is suited to cats which are thirsty, like salty foods and dislike fuss or attention.
SUGGESTED DOSE: Nat mur 30, twice daily for 10 days.

Bacillinum (Tuberculous lung)
This remedy is indicated in cases of ringworm. If Bacillinum fails to produce a cure, other remedies such as Sepia, Tellurium or Chrysarobinum can be tried.

SUGGESTED DOSE: Bacillinum 200, 1 dose weekly for 6 weeks.

Staphysagria (Stavesacre)
Classical miliary eczema with sparse hair over the pelvic area and small scabs over the back and around the neck often responds, where the condition arises shortly after neutering.
SUGGESTED DOSE: Staphysagria, 30 twice daily for 5 days.

Folliculinum (Corpus luteum)
This is another important remedy for miliary eczema in both male and female cats. Hormonal alopecia in spayed females may also respond.
SUGGESTED DOSE: Folliculinum 30, once daily for 1 month.

HERBAL REMEDIES

INTERNAL
Seaweed (Kelp, *Fucus vesiculosus*)
Seaweed encourages hair growth and improves skin pigmentation. Daily doses help keep the coat and skin in excellent condition.

Nettles (*Urtica dioica*)
Used on a daily basis, Nettles act as a general tonic for the coat, making the hair shine and removing scurf. Along with Seaweed, they can treat eczema (especially in nervous cats) and other skin problems.

Evening Primrose (*Oenothera biennis*)

The oil extracted from the seeds of this plant are rich in essential fatty acids. Deficiency can result in a poor coat and skin conditions such as miliary eczema. Daily dosing for several weeks may be needed before any improvement is noted.

EXTERNAL
Calendula (Marigold)
Diluted Calendula lotion is excellent for cleaning, bathing and soothing sore areas, scabs and rashes. The cream can be used on dry areas.

Myrrh (*Commiphora molmol*)
Used with other treatments, Myrrh tincture applied locally

will help clear up ringworm, particularly when combined 50:50 with Calendula lotion.

Pyrethrum (*Chrysanthemum cinerariifolium*)
Brush the dried, powdered flowers into the coat weekly to deter fleas. Groom the coat thoroughly 1 hour later to help remove any fleas. Pyrethrum is safe even for young kittens.

Lavender (*Lavandula officinalis*)
A little essential oil brushed into the coat weekly will deter fleas. Oils of Eucalyptus, Lemon, Rosemary, Citronella and Geranium can be used in the same way.

towelling at the bottom of the sink to stop your cat slipping and put a collar (not a flea collar) on it to give you a grip.

Gently place your cat in the water, talking to it reassuringly and wet its fur by carefully pouring water over it. Lather it up with shampoo, avoiding the head and ears completely, and rinse throughly. Wrap your cat in a towel (it may want to shake itself partly dry) and dry it as best you can. Then wrap another, dry towel around it and keep it in a warm draught-free room until it is completely dry. Some cats will tolerate being dried by a hairdryer but others will prefer sitting in front of the fire.

Shampoos which will kill fleas and other ectoparasites are available but do ensure that any shampoo used is suitable for cats. Fleas are tiny, reddish-brown insects but their droppings — tiny black specks in the fur which turn red when damp — are usually more visible. If you notice your cat scratching there may already be a large number present. They must be dealt with immediately as fleas can transmit serious diseases between cats and have a debilitating effect as they feed on a cat's blood. While scratching, a cat can injure itself, or set up an allergic reaction, leading to eczema. If it swallows a flea, a cat may develop tapeworm. In the western USA, bubonic plague is still found among rodents — if a hunting cat picks up infected fleas which then bite a human, death can result.

Cat fleas prefer cats to a human or a dog, but they don't spend all their life cycle in cat fur. They lay their eggs (as many as 300 a month each) in carpeting and furnishings. So when treating a flea infestation, it is important to treat the cat's environment as well as the cat. Environment anti-flea sprays kill fleas in the area and one containing methoprene also inhibits the development of the flea larvae, preventing them from becoming adults. Read the instructions carefully, take care never to use an environment spray on an animal and remove ornamental fish as well as food and water bowls from the room, as they may be poisoned.

Some people find that the safest flea control is the traditional fine-tooth comb. Fleas are extremely tough and impossible to squash so, if you use this method, be sure to throw them into the fire or a bowl of water. To prevent the fleas skating across the top of the water and climbing up the sides of the bowl add some washing-up liquid to break the surface tension. Some owners say that by combing their cats thoroughly every day they never have fleas.

There are a wide variety of chemical flea sprays and powders designed for use directly on the cat. Aerosol sprays are harmful to the environment, and many cats do not like them, so you may prefer to use a natural alternative or a flea powder which can be puffed on and brushed out. Brushes with hollow spikes are also available which, when filled with flea powder, dispense it through a cat's fur while it is being brushed. If you do use a spray, safeguard your health and your cat's by spraying outdoors and downwind and wear rubber gloves, especially if you are treating several cats. Spray in a short, sharp burst along the cat's back and belly — the fleas will die as they move around the cat. Avoid spraying your cat within several days of worming, as the two treatments together can prove toxic. It is also essential to check with your vet before using any flea products on kittens as some products are not suitable for them.

Herbal remedies are now available — as flea collars, powders and shampoos — and you can also make your own concoctions (see panel, 89). Bear in mind, however, that herbal products repel fleas without killing them.

Fleas hop from cat to cat. One new treatment involves applying a liquid to the cat's neck that enters the bloodstream and kills any fleas present. Tablets are also available from vets.

An ultrasonic device, said to emit the vibrations that repel fleas, is also available, and is worn attached to a cat's collar or placed in its bed. Another new device consists of a sticky pad under an electric bulb. The pad is saturated in flea pheromones and, once attracted, fleas are trapped by the adhesive.

Conventional flea collars work by releasing powdered insecticide gradually over a cat's coat. This takes time, so it will be a couple of days before it has an effect. If you use a collar replace it every two or three months, and check your cat's neck daily to ensure there is no inflammation. Take the collar off if it (or the cat) gets wet and do not replace until dry. Ideally, take the collar off when your cat goes outside. A survey carried out in the 1970s by the Royal Society for Prevention of Cruelty to Animals from a sample of 569 cats, showed that 161 were injured or died through wearing flea collars and 408 suffered from dermatitis round the neck.

Most flea products also dispose of lice, mites and ticks. Lice suck or bite a cat's skin and can debilitate it. Mites burrow into the skin, and ticks bite and hang on by their mouths — never pull them off or they will leave their teeth in the skin where an abscess may form. Vets treat ticks with chloroform, which encourages them to let go. A drop of castor oil may loosen the grip of a few ticks.

Another skin-related problem is ringworm — a fungal skin complaint, spread by airborne spores. It causes the fur to fall out in small, scaly, often circular patches. This needs veterinary treatment with antibiotics or an antifungal wash.

The Elderly Cat

A few cats live well into their twenties or even their thirties but most are not so fortunate. A cat of 10 can be considered a pensioner as its age equates to a person of about 60. Many cats stay naturally bright and active for many more years but there is a lot an owner can do to ensure a comfortable and healthy old age.

An elderly cat may slow down and become more interested in sleeping in a warm spot than in hunting or playing. Stiffening joints may mean that it can no longer groom itself so well, so comb it regularly, preferably daily, to prevent hairballs. Ageing muscles may no longer be able to expel hairballs so prevention is preferable to letting them build up.

The older cat has been a friend for years, so be a friend to him. You may need to be patient and help him with grooming, as well as ensuring his diet is correct.

Old bones enjoy warmth. If the sun does not provide it for your cat, make sure it has a warm, draught-free bed, ideally with a heating pad, to keep it comfortable and happy.

Pale-coloured cats may become grubby, so give a dry shampoo (see p. 88) before their fur becomes dirty.

Stiffness and slowness may mean that an older cat may not be able to reach its litter tray in time, so provide at least two. If you live in a house, stairs may slow your cat down, so have one tray upstairs and one down. Bending down to eat and drink may be difficult, too, so raise your cat's bowls on a platform or box or use one of the raised bowls made for the purpose.

Keep claws trimmed and check that the dewclaw (the claw a little way up the back of the leg) is not overgrowing into a curve which can pierce the flesh. You can trim this yourself as long as it is apparent which end to cut. If you are unsure, ask your vet to do it.

Some cats find their eyesight deteriorates and prefer to stay indoors. A degree of deafness is also not uncommon in later years and those who lose their hearing completely may miaow silently or howl very loudly. Other senses often develop to compensate. Some cats, for example, use traffic vibration to decide whether or not to cross the road. However, a deaf cat is safer kept indoors, as long as it does not object.

Older cats hate a change of routine, so try to keep to feeding times, grooming times and so on as far as possible. When you go on holiday, an older cat may find a cattery very stressful, so it may be better to ask a neighbour to come in and feed it while you are away.

Elderly cats cannot be expected to spend the

night outdoors. They feel the cold more than younger cats and need a warm bed, ideally with a pet-bed heater in it. If their favourite sleeping place is above floor level, put a chair underneath to help it climb up and down.

Kidney disease is the scourge of the older cat – as many as three-quarters of cats over the age of ten are at risk. If diagnosed early, it can be controlled by diet, allowing the cat a long and happy life. The first symptom is increased water consumption. The best treatment is less, but higher quality protein in the diet. Add some cooked potato, rice, pasta or bread to your cat's food – many cats adore pasta or rice, especially if cooked in a meaty stock. Mix in a little bran, too, if your cat seems constipated – a common problem in old age. There are now proprietary diets available from vets, specially formulated for older cats.

Many owners are surprised to learn that cats – especially older cats – can suffer from Diabetes Mellitus. If a cat eats and drinks more, yet loses weight, diabetes may be the reason. This is often treated with daily insulin injections, which owners learn to administer, but in some cases can be controlled by diet alone. There are also various homoeopathic remedies available that help to beat diabetes.

Gingivitis is fairly common among mature cats if they have not had regular tooth-cleaning sessions. Tartar accumulates on the teeth, causing inflammation of the gums, which can be seen as a red line between tooth and gums. If your cat appears to have mouth pain and is eating less, if it dribbles and has bad breath or if there is blood in the saliva, this may be the cause. Take your cat to the vet for thorough tooth cleaning and, if necessary, to have a tooth removed. Both are carried out under anaesthetic.

It is advisable to take an older cat for a veterinary check-up twice a year so that any problems can be picked up in their early stages. Failing that, see a vet *at least* once a year, to continue your cat's inoculations. If it finds travelling to the vet's surgery distressing, ask your vet to make a housecall. It costs a little more, but can be well worth it, especially if you have several cats.

Cancer in cats and dogs is more common than in humans. The danger years are between 10 and 15, with the incidence decreasing after that age. The treatment involves chemotherapy, radiotherapy or surgery.

Hyperthyroidism, the most common illness of older cats, occurs when the thyroid gland produces too much hormone, speeding up the metabolism. The symptoms include loss of weight, increased appetite and thirst, hyperactivity or aggression. Treatment is by chemotherapy, radiotherapy or surgery. Cats often respond well to homoeopathic treatment of this condition. Always bear in mind that any of these illnesses, although serious, will not prove fatal for your pet as long as veterinary attention is sought in time.

Some owners genuinely fear taking their older cats to the vet if they appear to be in pain, in case they are advised to have the cat put down. The truth is, however, that vets are in the business of saving lives and improving health. No vet will ever suggest putting a cat down unless it is in so much pain or the quality of its life has deteriorated to such an extent that it would be cruel to keep it alive. Unfortunately, this unwarranted fear can mean that some cats with minor but painful ailments do not receive the treatment which would make their golden years significantly more comfortable.

Contact your vet if your cat loses weight, eats and drinks more than usual, eats or drinks less than usual, or if you notice any change in its behaviour.

What if a cat is seriously ill or injured and living in constant pain? The owner may then have to consider discussing euthanasia with their vet. If a vet recommends euthanasia, the advice is not given lightly, and is likely to be the most humane option. It is easy to be so swayed by affection for a cat that we want to keep it alive at all costs. However, the finest expression of love is to allow ourselves to let go when the time has come.

If a cat is to be put down, it receives the same injection of barbiturates as for an anaesthetic but in a larger dose. Within 5 seconds the cat will be pleasantly drowsy. Within 10 seconds it will be asleep and within 15 seconds the cat will have passed on.

What it leaves behind is a devastated owner or family, grieving companion cats or dogs and many memories. It is natural to grieve over the death of a much-loved cat and it is important to do so. Repressing feelings means that scars never heal. Many people feel guilty and wonder what they could have done to prevent it. Usually the answer is – nothing. Accept that guilt is a part of the grieving process and do not let it weigh you down. Your cat was a friend, confidante and companion for many years. Allow yourself the time to grieve – it could be six months or more before you can think of your cat without that aching pang – but acceptance comes in time.

It is not always helpful advice to 'replace' your cat with another immediately. It is impossible to replace an individual, you can only acquire another. Those who are anxious to have a cat around the house need to be aware that a new companion will be totally different from the old.

Cat Massage

Massaging a cat is an extension of stroking. A cat's first experience of stroking is as a kitten, when its mother uses her rough tongue all along its body to clean and quieten it. Cat massage will help reduce stress, as well as relieving the effects of some ailments such as arthritis, cystitis and kidney problems.

When massaging your cat, always stroke in the direction of the fur. Never use oil on your cat and do not attempt to stroke the backbone or windpipe.

With your cat on your lap, stroke it until it relaxes. Using small circular movements gently massage along its sides with the fingertips, starting at the neck and working towards the tail. Massage the back of the neck and with vertical movements stroke the front of the neck. Stroke the legs between your fingers in a downward movement and massage between each pawpad. With circular movements, stroke the belly and finish by gently stroking from top to toe.

Massaging your cat is an extension of stroking and should be done very gently at first. Start massaging along its sides, working towards the tail but without touching the spine. Stroke the legs and massage between the pads of the paws. Finish with long strokes from head to toe.

Good health care keeps a cat contented. There are natural remedies for nearly all feline ailments.

A-Z OF COMMON AILMENTS

This reference section gives advice and natural remedies for a wide range of ailments. It is essential that you read the information on homoeopathy, herbalism and Bach flower remedies given on pages 9-16 before administering any of the following treatments. Also read 'Before You Start' on pages 16-17 for further essential preparatory information.

ABSCESSES

These arise from bite wounds that heal too quickly leaving bacteria trapped inside. The affected area swells, becomes painful and discoloured, as the body tries to expel the infected material.

HOMOEOPATHIC REMEDIES

Hepar sulph *(Calcium sulphide)*
This is a good remedy to use in instances when your cat has an abscess forming, which needs to be encouraged to burst. Use Hepar sulph in low potency to bring the abscess to a head.
SUGGESTED DOSE: Hepar sulph 3x, 3 times daily.

The same remedy can be used in high potency to prevent an abscess developing following a bite or to aid the resolution of one which is draining, particularly where the affected area is very painful.
SUGGESTED DOSE: Hepar sulph 200, twice daily.

Silicea (Flint)
Silicea is most helpful in treating chronic non-painful abscesses which fail to heal. Not only will it help eliminate the infection but also help to reabsorb any scar tissue.
SUGGESTED DOSE: Silicea 30, twice daily for up to 3 weeks.

HERBAL REMEDIES

INTERNAL
Garlic *(Allium sativum)*
Garlic has both antiseptic and antibacterial actions. It encourages drainage and assists healing.

Echinacea *(Echinacea angustifolia)*
This remedy has very effective antibacterial properties. Echinacea encourages the body to deal with the abscess infection

through its stimulation of the cat's immune system.

EXTERNAL
Calendula lotion (Marigold, *Calendula officinalis*)

Diluted Calendula lotion is ideal for cleaning away the discharge from a draining abscess. It will soothe the inflamed area and assist healing. Always make sure all the infected material has gone before you allow an abscess to heal over.

Slippery elm bark (*Ulmus fulva*)
Use a small amount of powdered bark mixed with boiling water to make a paste. When this has cooled down, apply a poultice to the affected area (see p. 15) to encourage the abscess to burst and help draw out the infected material.

ANAEMIA

Anaemia is essentially a lack of red blood cells. It can arise from blood loss or as a consequence of another condition, such as feline infectious anaemia, leukaemia, a liver problem or kidney disease.

HOMOEOPATHIC REMEDIES

Silicea (Flint)
This is the remedy of choice to use in cases where debility or malnutrition is the cause of the anaemia. Bear in mind, however, that an improved diet for the cat is also essential (see Chapter One).
SUGGESTED DOSE: Silicea 200, 3 times weekly for 6 to 8 weeks.

Arsenicum album (Arsenic trioxide)
Indicated in chronic anaemia or where haemorrhage has been the cause. The cat is easily exhausted, tends to be restless, seeks warmth and is thirsty for small amounts of water.
SUGGESTED DOSE: Ars alb 30, twice daily for 1 month.

Ferrum metallicum (Iron)
This is a more general remedy. It is particularly suitable to administer to weaker, lethargic animals. Cats that are often difficult to feed are also likely to respond well to treatment with Ferrum metallicum.
SUGGESTED DOSE: Ferrum met 30, once daily for 1 month.

HERBAL REMEDIES

INTERNAL
Seaweed (Kelp)
This is an excellent source of iron as well as trace elements, vitamins and minerals. Your cat will benefit from receiving it on a daily basis.

Nettles (*Urtica dioica*)
Nettles are rich in natural iron and in an easily assimilated form. In addition they strengthen the whole body.
Parsley, Watercress, Elderberry and Ginseng are also useful.

APPETITE

During and after illness or following surgery it is sometimes difficult to encourage cats to return to a normal diet. Some are naturally poor or fussy eaters. The following can help.

HOMOEOPATHIC REMEDIES

Nux vomica (Poison nut)
Nux vomica stimulates the liver and is a general tonic to the digestive system, increasing appetite. It is very useful after gastrointestinal upsets, especially

in combination with Carbo veg (Vegetable carbon).
SUGGESTED DOSE: Nux vomica 6, twice daily for 10 days.

Calc phos (Calcium phosphate)
In low potency Calc phos stimulates the appetite, assists digestion and helps the assimilation of food. It is a good convalescent remedy.
SUGGESTED DOSE: Calc phos 6x, twice daily.

HERBAL REMEDIES

INTERNAL
Condurango (*Marsdenia condurango*, Condor plant)
One of a number of herbs known as bitters which stimulate appetite. Condurango improves digestion and general health.

Gentian root (*Gentiana lutea*)
This is a well-known bitter, which stimulates the cat's digestion and appetite. It does this by increasing the flow of gastric juices and bile.

Chamomile (*Matricaria chamomilla*)
As well as improving the appetite, Chamomile acts as a gentle sedative, making it useful for nervous cats and during convalescence.

ARTHRITIS AND RHEUMATISM

Common in older cats, usually as a result of ageing or injury. Symptoms include stiffness, lameness and difficulty getting up or jumping.

HOMOEOPATHIC REMEDIES

Rhus tox (Poison ivy)
A well-known and extremely valuable remedy, but only where the symptoms match. Stiffness is evident after rest and worse for cold and damp. Movement eases the symptoms as does dry warm weather, but prolonged activity aggravates. Pain causes the cat to become restless and keep moving.
SUGGESTED DOSE: Rhus tox 6, twice daily to effect.

Bryonia alba (White bryony)
This remedy should be used in cases that contrast with those requiring Rhus tox in that the symptoms are worse for movement and better for rest. The cat prefers to lie still, tucking the affected limbs under the body. Warmth also tends to have the effect of aggravating the symptoms.
SUGGESTED DOSE: Bryonia 6, twice daily.

Actea rac or Cimicifuga (Black snake root)
When there is heaviness in the limbs and pain over the back, this remedy is extremely useful. Jumping also tends to be difficult for the cat and the limb movements are often jerky in such cases. Symptoms are generally worse in the morning and when the weather is cold.
SUGGESTED DOSE: Actea rac 6, twice daily.

HERBAL REMEDIES

INTERNAL
Nettles (*Urtica dioica*)
Their anti-inflammatory action reduces swelling and pain. They also promote urination helping to eliminate waste products that cause rheumatic pains. Add a pinch of dried Nettle powder to each meal.

Parsley (*Petroselinum crispum*)
The tonic effect of Parsley strengthens the whole of the cat's body and provides good supportive treatment in chronic cases. Good results can be obtained simply by adding a small pinch of chopped Parsley to food at each mealtime.

CAT FLU AND CHLAMYDIA

(See also Catarrh, below)

Cat flu is a common problem and caused by infection with either the feline rhinotracheitis virus and/or feline calici virus. Symptoms vary from fairly minor to quite severe, with respiratory problems, conjunctivitis, ulceration of the mouth and lethargy. Infection with *Chlamydia psittaci* can cause an illness which can resemble flu. In acute cases there is a severe conjunctivitis with red, grossly swollen conjunctival membranes and a pus-like discharge. Sneezing and a nasal discharge are often present. Apis mel (Honey bee) is a very valuable remedy in treating the eye symptoms. Details are included in the eye panel (see p. 78).

HOMOEOPATHIC REMEDIES

Aconite (Monkshood)
The remedy to use in the early stages when the cat is feverish and starts sneezing. Symptoms appear suddenly, often after exposure to cold, dry or very hot weather.
SUGGESTED DOSE: Aconite 30, every 2 hours.

Allium cepa (Red onion)
Another remedy best given early on and suited to minor infections. The nose has an unpleasant watery discharge

and the eyes water profusely and look sore.
SUGGESTED DOSE: Allium cepa 30, 3 times daily.

Arsenicum album (Arsenic trioxide)
Sores and scabs appear around the nostrils caused by a thin watery discharge from the nose. The eyes look sore and the discharge irritates the skin. In addition, the cat is restless, seeks warmth and drinks only small sips of water.

SUGGESTED DOSE: Ars alb 30, 3 times daily.

Merc corr (Corrosive sublimate of mercury)
A remedy to use in severe cases. There is a green discharge from the nostrils and eyes, which are very red and sore. The mouth is ulcerated causing profuse salivation. Merc corr is useful in treating infection with chlamydia.
SUGGESTED DOSE: Merc corr 30, 3 times daily.

HERBAL REMEDIES

INTERNAL
Garlic *(Allium sativum)*
Garlic has a special affinity for the respiratory tract. In addition to its antiviral and antibacterial actions, the volatile oils it contains help loosen and clear away the discharges.

Echinacea *(Echinacea angustifolia)*
A remedy to use in any infection due to its antimicrobial effect and wide safety margin. It is especially useful for treating infections involving the upper respiratory tract.

EXTERNAL
Euphrasia (Eyebright)
Put 1 or 2 drops of Euphrasia tincture in an eggcupful of cold boiled water. Use this to bathe your cat's sore eyes. You will probably need to do this 2 or 3 times every day.

CATARRH AND SINUSITIS

Upper respiratory tract infections, such as cat flu, frequently cause acute bouts of catarrh from which most cats recover. Chronic catarrh usually extends to the sinuses and often recurs even after antibiotic treatment.

HOMOEOPATHIC REMEDIES

Pulsatilla (Wind flower)
This remedy is indicated where the nasal discharge is yellow, creamy, bland and does not cause soreness around the nostrils. It loosens in fresh air. Pulsatilla suits cats that are affectionate and good-tempered.
SUGGESTED DOSE: Pulsatilla 6, twice daily.

Kali bich (Potassium bichromate)
In situations where Kali bich is needed the discharge from the cat's nose is thick, yellow, stringy and ropy, making the nostrils sore. The cat's sense of smell is diminished as the discharge plugs the nostrils.
SUGGESTED DOSE: Kali bich 6, twice daily.

Silicea (Flint)
Silicea is needed where the cat's nasal discharge is fairly loose, pale yellow and crusts around the nostrils, leaving sores underneath. This remedy tends to suit animals of a timid, quiet nature.
SUGGESTED DOSE: Silicea 30, twice daily.

HERBAL REMEDIES

INTERNAL
Golden seal (Hydrastis canadensis)
Hydrastis has a specific action on mucous membranes and is a good remedy to use in catarrhal problems. It reduces the

discharge and assists the body in removing the mucous.

Golden rod (Solidago virgauria)
A remedy useful in treating both acute and chronic cases. Not only does it have an anti-catarrhal

action but it is also anti-inflammatory.

Elderflower (Sambucus nigra)
Elderflower is good for catarrh and, particularly, sinusitis. It mixes well with Golden Rod.

CONSTIPATION

Causes are numerous including incorrect feeding, liver disease, dehydration, lack of muscle tone and damage to the pelvis, nerves and spine. The most common symptom is straining but this can also be associated with urinary problems or diarrhoea.

HOMOEOPATHIC REMEDIES

Nux vomica (Poison nut)
This remedy is needed where constipation arises from irregular movements within the intestines. There is much straining and either nothing is passed or only small amounts.
SUGGESTED DOSE: Nux vomica 30, twice daily for 7 days.

Opium (Poppy, Papaver somniferum)
Opium is useful where straining is absent and there is no desire to pass motions at all. The abdomen may be bloated or gassy and the patient often subdued.
SUGGESTED DOSE: Opium 30, twice daily for 5 days.

Lycopodium (Club moss)
Useful where constipation is due to liver disease. The abdomen is often bloated with gas and appetite is poor, the cat only eating small amounts at any one time.
SUGGESTED DOSE: Lycopodium 30, twice daily for 1 week.

HERBAL REMEDIES

INTERNAL
Rhubarb root (Rheum palmatum)
Use this mild purgative with

caution. It stimulates the gut, causing emptying, but can cause constipation. Use in acute but not chronic cases.

Yellow dock (Rumex crispus)
Use when liver disease is the cause. Dock stimulates the gut and the liver, promoting bile flow.

COUGHING AND BRONCHITIS

Respiratory problems are relatively common, especially bronchitis. Coughing, wheezing and laboured breathing are symptoms which may be seen. Severe coughing may cause retching, although a sore throat or a foreign body lodged in the throat can also cause this symptom.

HOMOEOPATHIC REMEDIES

Phosphorus (The metal)
The particular cough that responds well to phosphorus is dry and harsh, arising from a tickling sensation in the larynx. Symptoms are worse at night, on first settling down and in cold air.
SUGGESTED DOSE: Phosphorus 30, twice daily.

Bryonia alba (White bryony)
A hacking, dry, spasmodic cough causing the cat to sit up suggests Bryonia. The cough is worse at night, after eating or drinking and for moving around. Entering a warm room from outside can also induce coughing.
SUGGESTED DOSE: Bryonia 30, twice daily.

Drosera (Sundew)
Paroxysms of violent spasmodic coughing to the point of retching suggest using Drosera as a remedy. Sometimes these bouts of coughing can be so violent that the cat finds it very difficult to breathe.
SUGGESTED DOSE: Drosera 30, twice daily.

HERBAL REMEDIES

INTERNAL
Garlic (Allium sativum)
The volatile oils in Garlic are excreted through the respiratory tract making it invaluable for chest problems, particularly bronchitis. It can be given as a preventative remedy daily.

Coltsfoot (Tussilago farfara)
Coltsfoot is both an anti-spasmodic and expectorant with anti-inflammatory properties due to its zinc content. It is a good general remedy to use in treating cats suffering from irritating coughs and bronchitis.

White horehound (Marrubium vulgare)
This remedy combines well with Coltsfoot and is useful in treating bronchitis. It has both anti-spasmodic and expectorant properties, dilating the airways and helping to loosen mucous.

CYSTITIS

This is a common problem, particularly in older females. Your cat may keep straining, trying to pass urine. Only small amounts are passed at a time, often with blood and jelly-like mucous. Repeated attacks may need a change in diet in consultation with your vet.

HOMOEOPATHIC REMEDIES

Cantharis (Spanish fly)
Frequent and severe straining accompanied by general irritability characterize this remedy. Only small amounts of bloody urine are passed, usually containing mucous. Where symptoms are similar but less intense try Juniper (Juniper berries).
SUGGESTED DOSE: Cantharis 30 or Juniper 30, 3 times daily.

Apis mel (Honey bee)
With Apis the bladder is painful and bloody urine is passed drop by drop, sometimes in dribbles. The cat seeks cold places to sit and refuses to drink.
SUGGESTED DOSE: Apis mel 30, 3 times daily.

Equisetum (Scouring rush)
Indicated where large quantities of urine are passed drop by drop and pain is evident at the end of urination. Passage of urine does not relieve the discomfort. Incontinence in older cats may respond.
SUGGESTED DOSE: Equisetum 30.

HERBAL REMEDIES

INTERNAL
Buchu (Agathosma betulina)
Both soothing and healing to the urinary tract. Not only is Buchu valuable in treating chronic cystitis and incontinence but it also promotes urination and helps dissolve gravel.

Bearberry (Arctostaphylos uva-ursi)
Bearberry is very good for soothing and strengthening the urinary tract, having both astringent and antiseptic effects. It has a similar action to Buchu with which it combines well.

Parsley piert (Aphanes arvensis)
This little herb promotes a good flow of urine and is used principally to help remove gravel. In addition its protective demulcent action soothes the urinary tract where urination is painful.

DIARRHOEA

Causes are numerous including worms, bacterial and viral infections, food poisoning, food allergy, poor diet and liver or kidney disease. Mild cases clear with a period of starvation followed by a bland diet. Milk should be withheld.

HOMOEOPATHIC REMEDIES

Podophyllum (May apple)
This remedy is good for treating chronic diarrhoea in kittens if the following symptoms are present. The stools are watery, yellow, profuse and gush out with force. Gurgling noises in the abdomen and prolapse of the anus are useful guiding symptoms in such cases.
SUGGESTED DOSE: Podophyllum 30, twice daily.

Aloe (Aloe socotrina)
Aloe is an appropriate remedy in cases where the cat has a great urge to pass faeces that are jelly-like and can contain blood. Gas can also be present, which sometimes causes spluttering as the stools are passed. In addition, the anus is weak and loose motions may dribble out.
SUGGESTED DOSE: Aloe 30, twice daily.

Arsenicum album (Arsenic trioxide)
Arsenicum is indicated where there is extreme lethargy (prostation) and the stools are small, watery, bloody and foul smelling. The anus is red and sore and the cat is thirsty for small amounts. Symptoms are worse at night.
SUGGESTED DOSE: Arsenicum 30, 3 times daily.

HERBAL REMEDIES

INTERNAL
Garlic (Allium sativum)
Garlic disinfects the digestive tract and helps restore normal levels of bacteria in the gut by killing pathogenic bacteria.

Marshmallow root (Althaea officinalis)
Marshmallow is a demulcent (lubricant) and soothes the lining of the digestive tract where it is inflamed and assists in healing. It is of most use in treating long-standing cases of diarrhoea and colitis.

Slippery elm bark (Ulmus fulva)
This remedy can be combined with marshmallow in the treatment of diarrhoea and colitis. Not only does it have a soothing effect, but it is also an astringent, toning the gut lining.

Barberry (Berberis vulgaris)
By stimulating the liver, Barberry promotes the flow of bile and other intestinal secretions which help the bowel return to normal.

FRACTURES

Whilst veterinary help is needed initially, a number of remedies are available to assist healing. Always give Arnica to help with bruising.

HOMOEOPATHIC REMEDIES

Symphytum (Comfrey, Knitbone)
Comfrey stimulates the healing of fractures. It is most valuable in cases of non-union where healing is unusually slow.
SUGGESTED DOSE: Symphytum 6, twice daily.

Ruta grav (Rue)
Ruta acts upon periosteum (the outer layer of bone) and cartilage and is one of the main remedies for treating bone injuries.
SUGGESTED DOSE: Ruta 30, once daily.

Calc phos (Calcium phosphate)
This remedy helps to regulate calcium metabolism and will assist in the healing of fractures. Like Symphytum, it is particularly useful in cases where non-union is evident.

HERBAL REMEDIES

INTERNAL
Comfrey (*Symphytum officinale*)
Taken internally Comfrey will stimulate fractures to heal, hence its old fashioned name – Knitbone. It can also be used

externally, over the fracture site, as a poultice, compress or cream.

EXTERNAL
Mouse ear (*Pilosella officinarum*)

Applied externally as a poultice, Mouse ear is a good remedy for speeding up fracture union. Horsetail (*Equisetum arvense*) can also be used in this way as an alternative.

FUS (feline urological syndrome)
(See also Cystitis, pp. 102-103)

Caused by the formation of gravel in the urine which can plug the urethra so that urine cannot be passed. The acute condition is an emergency needing immediate veterinary attention. Where the condition is chronic a number of remedies are useful in helping to dissolve gravel after consulting the vet.

HOMOEOPATHIC REMEDIES

Berberis (Barberry)
Very useful in helping to dissolve gravel and in preventing cystitis, which often accompanies the condition. Tenderness over the lumbar region is a good guiding symptom to its use.
SUGGESTED DOSE: Berberis 6, twice daily for 1 month.

Hydrangea (*Hydrangea arborescens*, Seven-barks)
A remedy specifically for helping to dissolve gravel and calculi especially where there is mucous in the urine. It can be used as a preventative.
SUGGESTED DOSE: Hydrangea 3x, twice daily.

HERBAL REMEDIES

INTERNAL
Gravel root *(Eupatorium purpureum)*
Promotes urine flow, helping flush out minerals that make up

gravel and dissolve gravel.

Stone root *(Collinsonia canadensis)*
This is a useful remedy in

preventing the condition, which can be given daily to cats likely to develop problems. In addition, Stone root can be effectively combined with Gravel root.

JAUNDICE
(See also Liver Problems, p. 106)

Arises when the liver becomes congested or the flow of bile is obstructed. Bile pigments can no longer be excreted and build up in the body, imparting the characteristic yellow colour to the skin and mucous membranes.

HOMOEOPATHIC REMEDIES

Chelidonium majus (Greater celandine)
This remedy has a specific action on the liver. It is needed where certain symptoms are present: the faeces are pale, bright yellow or clay coloured. In addition, diarrhoea may alternate with constipation and the liver is

usually enlarged.
SUGGESTED DOSE: Chelidonium 30, twice daily for 10 days.

Carduus marianus (St Mary's thistle)
Another remedy with its action centred on the liver. Faeces are hard and clay coloured. A good

remedy for treating jaundice, especially where the liver is engorged.
SUGGESTED DOSE: Carduus 30, twice daily for 2 weeks.
There are also herbal remedies that are useful in treating jaundice. These are listed in Liver Problems (see p. 106).

KIDNEY PROBLEMS

One of the most frequently encountered problems among older cats. The signs include gradual weight loss, increased thirst, lethargy, poor appetite, bad breath and other mouth problems. A large number of cases respond to herbal or homoeopathic treatment in conjunction with a suitable diet.

HOMOEOPATHIC REMEDIES

Natrum muriaticum (Sodium chloride, common salt)
This is a suitable remedy for cats which prefer to be left alone and do not like fuss. Increased thirst and frequency of urination are keynotes, usually accompanied by a good appetite, although the cat loses weight. Dehydration is evident and the coat is unkempt and greasy in appearance.
SUGGESTED DOSE: Nat mur 30,

once daily for 2 weeks then 200 potency 3 times weekly for 1 month.

Arsenicum album (Arsenic trioxide)
Arsenicum is a valuable remedy where there is dehydration, thirst for small quantities of water and the coat is dry and flaky. Irritation of the skin may be apparent with symptoms worse

towards midnight. This remedy is suitable to give to cats of a nervous, restless nature.
SUGGESTED DOSE: Arsenicum 30, twice daily for 2 weeks.

Plumbum metallicum (Lead)
This remedy is useful when muscle wasting is the most prominent guiding symptom, leading to weakness and difficulty moving. Since lead also causes

constipation and anaemia, these symptoms may be apparent too.
SUGGESTED DOSE: Plumbum 6, twice daily.

Merc sol (Mercury, quicksilver) Useful in cases where there are mouth problems, principally gingivitis, bad breath and sali-

vation. Thirst is increased and urine tends to be dark brown.
SUGGESTED DOSE: Merc sol 30, twice daily for 2 weeks.

HERBAL REMEDIES

INTERNAL
Bearberry (Arctostaphylos uva-ursi)
This remedy has a specific action on the urinary tract and is one of the best herbs to treat kidney failure. It promotes urine flow and has a soothing antiseptic effect.

Buchu (Agathosma betulina)
Another remedy with diuretic and urinary antiseptic properties whose actions complement those of Bearberry. Kava kava (Piper methysticum) can be used as an alternative to Buchu as it has similar properties.

Herbs such as Dandelion and Barberry (see Liver Problems) which promote the flow of bile are also of value when treating your cat for kidney problems. This is because they assist in the excretion of toxins from the body.

LIVER PROBLEMS

(See also Jaundice, p. 105)

The liver is one of the major organs of the body and central to well-being. A tendency to keep eating grass, periodic vomiting, odd-coloured soft faeces, increased thirst and weight loss are a few of the symptoms associated with liver disease.

HOMOEOPATHIC REMEDIES

Nux vomica (Poison nut)
This remedy helps clear the liver of toxins and also acts as a tonic to the digestive system. Irritability, periodic vomiting and constipation point to its use.
SUGGESTED DOSE: Nux Vomica 30, twice daily for 5 days.

Phosphorus (The element)
Vomiting within an hour of eating, increased thirst, soft clay-

coloured faeces and bleeding gums suggest Phosphorus.
SUGGESTED DOSE: Phosphorus 30, once daily for 2 weeks.

Lycopodium (Club moss)
The remedy of choice for chronic liver congestion, where fluid may have collected in the abdomen. The liver region is tender and the abdomen bloated. Motions tend to be small and hard.

SUGGESTED DOSE: Lycopodium 30, once daily for 2 weeks.

Berberis vulgaris (Barberry)
Berberis is useful where there is vomiting early in the morning and poor appetite, accompanied by pain over the back and weakness of the legs. It is also a good kidney remedy.
SUGGESTED DOSE: Berberis 30, once daily for 2 weeks.

HERBAL REMEDIES

INTERNAL
Dandelion (Taraxacum officinale)
A good tonic, useful in inflammation of the liver and when there is congestion and jaundice. It stimulates bile production and is a potent

diuretic, promoting urine flow, cleansing the body of impurities.

Barberry (Berberis vulgaris)
Promotes bile flow and helps restore liver function, particularly where jaundice is present. It is also a mild laxative.

Fringetree (Chionanthus virginicus)
Fringetree bark combines well with Barberry and is a safe and effective remedy to apply to all liver problems. It is especially useful where jaundice has developed.

SPRAINS AND STRAINS

The most obvious signs are lameness, stiffness or difficulty moving. Affected joints, muscles or ligaments are usually tender. Rest is important in addition to any treatment given.

HOMOEOPATHIC REMEDIES

Arnica montana (Leopard's bane)
Give 2 or 3 doses of Arnica initially to reduce inflammation and bruising. Where soreness is very prominent give Bellis Perennis (The Daisy) instead.
SUGGESTED DOSE: Arnica 30, 2 or 3 tablets 1 hour apart.

Rhus tox (Poison ivy)
Rhus tox has a special affinity for tendons and joints. The most important guiding symptoms are stiffness and pain, worse for resting and eased with movement.
SUGGESTED DOSE: Rhus tox 6, 3 times daily.

Ruta graveolens (Rue)
Ruta is an excellent remedy for treating strains. It acts particularly on ligaments and tendons where they attach to bone.
SUGGESTED DOSE: Ruta 30, 3 times daily.

HERBAL REMEDIES

EXTERNAL
Arnica
Provided that the cat's skin is not broken as a result of the sprain or strain, Arnica lotion can be applied locally as a compress (see p. 15). Do not use this remedy on damaged skin: use Witch hazel instead. Arnica or Rhus tox ointment can also be of value in treating sprains and strains.

Rosemary (*Rosmarinus officinalis*) Rosemary can ease muscle pain when applied as an infusion locally. Use 1 teaspoon of dried herb added to a cup of boiling water. Use when cool. The essential oil can also be used. Add 1 or 2 drops to 10 ml of Hazelnut oil and work gently into the region twice daily. Lavender oil is a good alternative.

VOMITING

All cats vomit occasionally, particularly after eating grass. This is normal: your cat is eliminating toxins from its body. Chronic periodic vomiting can be caused by hairballs, foreign bodies, liver or kidney disease. More acute bouts may be due to bacterial or viral infections, a change in diet or eating something unsavoury. A period of starvation is nearly always necessary with any treatment.

HOMOEOPATHIC REMEDIES

Phosphorus (The element)
A remedy associated with thirst where the patient drinks larger amounts of water, which are likely to be vomited back some minutes later. If food is vomited, large mouthfuls are brought back each time. A good remedy for post-operative vomiting.
SUGGESTED DOSE: Phosphorus 30, 3 times daily.

Nux vomica (Poison nut)
Useful where vomiting has been caused by eating rich food or a change in diet. Both retching and vomiting are violent. The symptoms are worse in the morning and after eating, but better for having been sick.

SUGGESTED DOSE: Nux vomica 30, 3 times daily.

Ipecacuanha (Ipecac-root)
This is the remedy of choice in cases of repeated vomiting, where the cat's symptoms do not abate for having been sick. In such instances, the cat will exhibit profuse salivation and a lack of thirst. In addition, the vomit may contain traces of blood.
SUGGESTED DOSE: Ipecac 30, 3 times daily.

HERBAL REMEDIES

INTERNAL
Peppermint *(Mentha piperita)*
The oils in Peppermint will soothe the lining of the stomach, reducing the feeling of nausea and the urge to vomit. It also stimulates the digestion and reduces gas.

Chamomile *(Matricaria chamomilla)*
Chamomile has a gentle sedative effect in addition to soothing action on the digestive system. It is also anti-inflammatory and can relieve the discomfort associated with gastritis.

Marshmallow root *(Althaea officinalis)*
Has a very soothing effect and is valuable in treating cases of chronic vomiting where the stomach lining is inflamed and irritated.

First Aid

BITES

Read 'Before You Start', pp. 16-17

Cats frequently get bitten during fights. First aid treatment can often prevent infection and abscess formation. Snake bites can also be a problem in some parts of the world.

HOMOEOPATHIC REMEDIES

Ledum palustre (Marsh tea)
This is the remedy to use for all puncture wounds produced by sharp objects or bites, particularly if the area around the wound becomes discoloured.
SUGGESTED DOSE: Ledum 30, 2 or 3 doses 1 hour apart.

Hepar sulph (Calcium sulphide)
Given in high potency, the use of this remedy can prevent wounds becoming infected. It is particularly good where the wound is painful and very sensitive to touch.
SUGGESTED DOSE: Hepar sulph 200, twice daily.

Crotalus horridus (Rattlesnake venom)
Use this remedy where the wound has become swollen, painful and discoloured. There may be seepage of dark blood and the cat seems lethargic. A good remedy (as might be expected) for treating snake bites.
SUGGESTED DOSE: Crotalus 30, every 2 hours.

Echinacea (Cone flower)
Echinacea is indicated when a septic state has arisen after the cat has been bitten. The discharge from the wound is foul and the cat has a fever. Snake bites, including those where gangrene threatens, can also respond to Echinacea.
SUGGESTED DOSE: Echinacea 6x, 4 times daily.

HERBAL REMEDIES

INTERNAL
Echinacea (*Echinacea angustifolia*)
This remedy helps the body to cope with infections by stimulating the cat's immune system. It also has beneficial anti-bacterial properties.

EXTERNAL
Witch hazel (*Hamamelis virginiana*)
Distilled Witch hazel has an astringent action that helps arrest bleeding and also reduces bruising and inflammation. Apply directly using cotton wool.

Calendula (Marigold)
This is a very effective remedy for bites. Diluted Calendula lotion can be used to bathe bite wounds. As well as stimulating healing it also helps reduce pain and inflammation.

BRUISING

Read 'Before You Start', pp. 16-17.

HOMOEOPATHIC REMEDIES

Arnica (Leopard's bane)
Commonly known as the 'fall herb', Arnica is one of the best remedies to use in any injury or potential injury (such as surgery). If it is given promptly to your cat it can prevent the development of bruising, reduce pain and limit bleeding.
SUGGESTED DOSE: Arnica 30, every 3 to 4 hours.

Bellis perennis (Daisy)
This is the best remedy to use when the bruising is deep and severe. It is particularly good in treating bruising to the pelvic area and after the cat has undergone major surgery.
SUGGESTED DOSE: Bellis 30, 3 times daily.

HERBAL REMEDIES

EXTERNAL
Arnica (Arnica montana)
Provided the skin is unbroken, dilute Arnica lotion can be applied as a compress. Alternatively apply Arnica cream.

Rue (Ruta graveolens)
Ruta ointment can be applied directly where there is bruising over bone.

Witch hazel (Hamamelis virginiana)
Distilled Witch hazel can be applied locally as a compress to limit bruising. In addition its astringent action allows it to be used where skin is broken to arrest bleeding.

Elder (Sambucus nigra) and Agrimony (Agrimonia eupatoria)
Make an ointment by mixing together 1 part dried Agrimony, 3 parts fresh Elder leaves and 6 parts Vaseline. Heat the mixture until the leaves are crisp, then strain it. This makes an excellent application to heal bruises on your cat's body.

BURNS AND SCALDS
Read 'Before You Start', pp. 16-17.

Run cold water over the affected area for several minutes to minimize the effects of the burn.

HOMOEOPATHIC REMEDIES

Arnica montana (Leopard's bane)
A single dose of Arnica 200 can be given as soon as possible after the accident. This will work effectively to reduce pain, limit swelling and allay shock.

Cantharis (Spanish Fly)
Prompt use reduces inflammation and soreness, relieving discomfort and pain.
SUGGESTED DOSE: Cantharis 30, 3 times daily.

Urtica urens (Stinging nettle)
Urtica can be used as an alternative to Cantharis, reducing pain and limiting swelling. Ointments containing Arnica and Urtica can be used externally.

HERBAL REMEDIES

EXTERNAL
Calendula (Marigold) and Hypericum (St John's Wort)
The soothing, healing action of Calendula combined with the pain-relieving qualities of Hypericum is ideal for treating burnt skin. Apply either the diluted tincture or Hypericum/Calendula ointment 3 or 4 times daily.

Aloe (Aloe vera)
Aloe gel or the fresh juice from a cut leaf can be applied locally to minor burns to reduce pain and provide a protective layer against infection.

BACH FLOWER REMEDY
Rescue Remedy
Burnt animals are often very frightened so a few drops of prepared rescue remedy on the tongue will have a calming effect and reduce shock.

HAEMORRHAGE (BLEEDING)

Read 'Before You Start', pp. 16-17.

Prompt veterinary attention is always needed to stop serious bleeding. As a first-aid measure, hold a pressure pad over the wound. The following can also be very helpful.

HOMOEOPATHIC REMEDIES

In most cases dose with the appropriate remedy, 30c potency every 15 minutes.

Arnica montana (Leopard's bane)
Haemorrhage with bright-red blood arising from injury, whatever the cause, suggests Arnica montana.

Millefolium (Yarrow)
The indications for using this remedy on your cat are almost exactly the same as those that call for Arnica montana. Millefolium is also good for

treating nosebleeds and is useful in cases of poisoning with an anti-coagulant rat poison.

Ferrum phos (Iron phosphate)
This is a good remedy to use when the haemorrhage is of bright-red blood. Clots are also present, which are another helpful indication of when this particular treatment should be used.

Hamamelis (Witch hazel)
Useful where the blood is dark, does not clot easily and oozes persistently.

Crotalus horridus (Rattlesnake venom)
A good remedy to use in infected bleeding wounds where the blood is dark and watery, almost black. Clotting does not occur easily, the blood remaining fluid. Lachesis (Bushmaster venom) is a good alternative.

Strontia (Strontium carbonate)
This is an excellent post-operative remedy. It is particularly useful for wounds that slowly seep blood. It also helps combat shock.

HERBAL REMEDIES

EXTERNAL
Distilled Witch hazel (*Hamamelis virginiana*)
An astringent that will help arrest bleeding in any situation. Apply using moistened cotton wool.

Lady's mantle (*Alchemilla vulgaris*)
An infusion of Lady's mantle can be applied to the wound as a compress to help stop your cat bleeding. This remedy also has a

valuable anti-inflammatory effect. There are various other astringent remedies that can be used in a similar way. These include Yarrow, Plantain and Oak bark.

SHOCK AND COLLAPSE

Read 'Before You Start', pp. 16-17.

Although a seriously ill or injured cat needs immediate veterinary care, an appropriate remedy given promptly can mean the difference between life and death.

HOMOEOPATHIC REMEDIES

Arnica montana (Leopard's bane)
One of the best remedies to give in any case of shock or trauma such as a road accident. Given quickly it will help with both mental and physical aspects,

limiting the effects of the injury and helping to arrest any bleeding.
SUGGESTED DOSE: Arnica 30, every 15 minutes.

Aconite (Monkshood)
The principal remedy in cases of shock, sudden fright and anguish. Symptoms are restless, frantic and fearful behaviour and rapid breathing.

SUGGESTED DOSE: Aconite 30, every 15 minutes.

Carbo veg (Vegetable carbon)
A classic remedy for treating collapse. It is also known as 'the corpse reviver', which aptly describes its effect. In particular, it is good for circulatory collapse where the cat shows hunger for air. Symptoms include open mouth, blue tongue and icy cold-ness, nearing the point of death.
SUGGESTED DOSE: Carbo veg 200, every 15 minutes.

Cinchona officinalis (Peruvian bark, China)
This remedy is indicated in cases where collapse arises due to loss of body fluids, particularly blood (from haemorrhage) but also from the fluid lost in cases of repeated vomiting and prolonged diarrhoea. There are other tell-tale symptoms to look for as well: lethargy accompanies sunken eyes, weak pulse and tacky skin.
SUGGESTED DOSE: China 30, hourly.

Veratrum album (White hellebore)
Use Veratrum to treat a cat suffering from post-operative shock. Symptoms to look for are slow anaesthetic recovery and the patient will tend to be cold with a dry mouth and poor pulse. A similar picture presents in some cases of gasteroenteritis where the cat is thirsty but vomits straight after drinking and has watery diarrhoea.
SUGGESTED DOSE: Veratrum 30, every 30 minutes.

REMEDY

BACH FLOWER REMEDY
Rescue remedy
This is a mixture of five different flower remedies that cover the various mental states which arise from an accident or shock. Particularly the Star of Bethlehem helps with the physical and mental effects of shock and Rock Rose deals with terror and panic.

Give 2 drops of prepared remedy by mouth every 10 minutes. Where this is not possible, the undiluted stock remedy can be applied directly to the skin.

WASP AND BEE STINGS
Read 'Before You Start', pp. 16-17.

Cats often chase and play with wasps or bees and inevitably get stung. Local swelling around the sting is common but more severe allergic reactions sometimes occur. Symptoms include blotchy swellings, salivation, rapid breathing, vomiting or lethargy, occasionally collapse in severe cases.

HOMOEOPATHIC REMEDIES

Ledum palustre (Marsh tea)
Use Ledum for puncture wounds of any description including insect bites. If given promptly it will help antidote the effects of the venom. The mother tincture can be applied locally to wasp stings.
SUGGESTED DOSE: Ledum 200, 2 doses 30 minutes apart.

Apis mel (Honey bee)
This remedy can be of considerable help to a cat suffering from a sting in cases where the region of the bite swells rapidly and is very sore and painful in appearance.
SUGGESTED DOSE: Apis 30, every 30 minutes.

Urtica urens (Stinging nettle)
This remedy is needed where a wasp or bee sting has brought about considerable skin irritation, particularly in cases where there are raised urticarial blotches on your cat's skin. Urtica mother tincture can be applied directly to the site of bee stings.

HERBAL REMEDIES

EXTERNAL
Aloe (Aloe vera)
Where the area around the bite is sore, the fresh juice from a broken leaf can reduce the inflammation. Fresh Plantain *(Plantago major)* or Houseleek *(Sempervivum tectorum)* leaves can be used as alternatives.

Lavender and Eucalyptus Painful swellings can be reduced with a drop of the neat essential oil of either Lavender or Eucalyptus applied locally to the skin.

Wasp stings are alkaline and a little Thyme vinegar dabbed on to the bite can antidote the venom. Sodium bicarbonate dissolved in ice-cold water will help with bee stings which are acid.

WOUNDS AND CUTS
Read 'Before You Start', pp. 16-17.

HOMOEOPATHIC REMEDIES

Arnica montana (Leopard's bane) Give this remedy to your cat immediately following any injury. Not only is Arnica montana beneficial in reducing shock, it will also help to stop swelling, bruising and bleeding and to reduce the risk of infection. SUGGESTED DOSE: Arnica 30, hourly until relief.

Hypericum (St John's wort) Hypericum given promptly will reduce pain significantly. It is good for lacerations, grazes or where tissue has been crushed. The diluted tincture can be used externally in combination with Calendula. SUGGESTED DOSE: Hypericum 30, 2 or 3 times daily.

Hepar sulph (Calcium sulphide) Use this remedy to treat infected and suppurating wounds. Affected areas are painful to touch and sensitive to cold and draughts. Warmth will help to ease your cat's discomfort. SUGGESTED DOSE: Hepar sulph, 30 twice daily.

HERBAL REMEDIES

EXTERNAL
Woundwort *(Stachys palustris)* As its name suggests this herb is a renowned healer. Use the diluted tincture to cleanse wounds or to make a compress to cover the affected area (see p. 15).

Self-heal *(Prunella vulgaris)* Self-heal helps cuts and wounds heal cleanly. The fresh leaves can be applied directly. A poultice or compress can be used instead.

Comfrey *(Symphytum officinale)* Speeds up wound healing and at the same time limits the formation of scar tissue. Use either a compress or as a poultice (see p. 15). It is good for deep wounds, but be sure that they are clean before they close over completely.

BACH FLOWER REMEDY
Rescue Remedy
The cream can be applied locally to any injured area. Use prepared stock remedy internally if shock or panic is evident.

Preventative Health Check-list

Daily tasks

- Feed your cat at least twice daily. Several small meals keep the urine pH balance steadier than one big meal, which is important for your cat's health.
- Give your cat fresh water to drink, as it helps to flush any impurities out of the body.
- Clean out litter trays – twice a day, more often if possible. Encourage your cat to urinate frequently, otherwise urological problems may occur.
- Play with your cat. A short exercise session tones up muscles, keeps the cardio-vascular system healthy and strengthens the bond between cat and owner.
- If your cat is longhaired, especially if a Persian, comb its fur thoroughly to prevent hairballs forming.

Weekly tasks

- If your cat is shorthaired, give it a weekly grooming session. Brush a little essential oil of lavendar, eucalyptus, lemon, rosemary, citronella or geranium into the coat every week to prevent fleas. Give all cats one oily meal per week, such as tinned mackerel or sardines in oil, to help ease the passage of any swallowed fur.
- If you are concerned about your furniture or skin, give your cat's claws a trim every one to two weeks.

Monthly tasks

- Use a hairball treatment, if necessary.
- If the weather has been warm, you may need to use flea treatments monthly, or even more often. Fleas carry feline diseases and can cause allergies.
- Every two to three months, use a roundworm preparation, and when necessary, add a tapeworm preparation. Worms cause illhealth and liver and lung damage.

Yearly

Carry out necessary inoculations to prevent serious or fatal illnesses. A combined inoculation against feline infectious enteritis and feline respiratory disease can be given at nine weeks and another at twelve weeks. Thereafter, annual boosters keep up immunity. Rabies vaccinations can be given at three months, thereafter annual boosters are given. Currently only the UK, Australia, New Zealand, Hawaii and parts of Scandinavia are rabies-free. A vaccination against feline leukaemia virus can be given to a kitten at nine weeks with a second two to three weeks later. Annual boosters should be given and this can be done at the same time as other protective immunization.

Carrying Your Cat

(See also Remedies for Stressful Situations, p. 55)

Do you need to take your cat to the vet? Does it struggle, splay out all four legs like a vertical-take-off aircraft, go rigid so that there is no way you can get it into the carrier, and generally have nothing to do with the idea? Are you surprised?

The vast majority of cats never see a carrier until there is something wrong with them. They are put into a carrier, taken to the veterinary surgery and jabbed in unmentionable places with a variety of painful instruments. Little wonder then that the appearance of the cat carrier prompts the speedy disappearance of the cat.

You can break out of this vicious circle the day you pick up your new cat or kitten by bringing in the carrier, opening it for inspection and sprinkling a little catnip or some dried food into it. While you are chatting with your cat's previous owner, it can investigate the carrier at its leisure and, if it finds something appealing inside, it will be less averse to spending a few minutes in it.

Do not get into the habit of getting out the carrier only when something unpleasant has to be done. If you have the time, take out the carrier occasionally, place your cat in it and drive it round the block. Bring it back home, give it praise and something nice to eat as a treat. Your cat will then associate the carrier with enjoyable experiences and will come to expect a pleasant treat after its journey. And if you do have to take your cat to the vet, something nice to eat afterwards will soon rid its memory of any unpleasantness.

There is a wide variety of cat carriers available. The cardboard carrier is not recommended. Although cheap, it has two significant disadvantages. One, since the cat cannot see out it may become terrified and try to tear its way out (it may succeed). Two, a frightened cat may urinate. If it does so in a cardboard carrier, the cardboard will disintegrate and the cat will escape.

It is better to pay a little extra initially and look on a cat carrier as a lifelong investment,

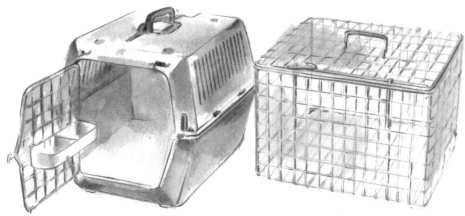

Two popular types of cat carrier that are very easy to clean. The one on the left comes apart for easy storage. The one on the right opens at the top, which helps the removal of a reluctant cat.

Always hold a cat gently but securely around the chest or back with your hand under its rear.

as a good carrier will last the lifetime of a cat and beyond. The ideal carrier is easily cleaned and has a top opening; it is almost impossible to remove a reluctant cat from a front-opening carrier.

Carriers made entirely of robust plastic-covered wire with a secure fastener are ideal. Although they may be slightly more expensive, they will probably last longer than other types. Some carriers have a plastic-covered wire front with a plastic or glass-fibre body. These are not as strong although they may have the benefit of coming apart for stacking or folding flat. However, at least one cat has gone missing on a subway system when the

train door closed on this type of carrier, squashing it and releasing its occupant.

Wickerwork carriers look attractive but are almost impossible to disinfect. Besides which, when you carry them, they ladder your stockings or damage your trousers.

Whatever type of carrier you choose, do not feed your cat or kitten within a few hours of the journey as it may become travel sick. Drive slowly, taking corners smoothly and, if possible, have someone other than the driver available to talk soothingly to the cat. Many cats will shriek throughout a car journey while others are silent; this is a matter of temperament, however, and does not necessarily mean that the noisy cats are more upset than the silent ones.

When lifting your cat to place it in a carrier (or for any other reason), always speak soothingly to it. Place one hand around the chest and the other under its rear and lift. Do not hold by the scruff of the neck, even with the other hand under the rear as this places too much strain on the neck muscles. Mother cats lift their kittens this way, but as they do not have hands, they have no other option. Kittens are often hurt this way when they are banged on the ground or on furniture; also the skin of the neck is frequently broken by the mother's teeth. Mother cats move their kittens only when necessary so it is best to follow their example.

If you can make a cat carrier a pleasant place to be and your cat jumps in voluntarily, this is very much better than having to persuade it to enter.

SUPPLIERS (UK)

HOMOEOPATHIC REMEDIES
Ainsworths Homoeopathic Pharmacy,
38 New Cavendish Street,
London W1M 7LH
(071) 935 5330

Ainsworths Homoeopathic Pharmacy,
40-44 High Street,
Caterham, Surrey CR3 5UB
(0883) 340332

Galen Homoeopathics,
Lewell Mill,
West Stafford, Dorchester,
Dorset DT2 8AN
(0305) 263996

Helios Homoeopathic Pharmacy,
92 Camden Road,
Tunbridge Wells, Kent TN1 2QP
(0892) 536393

A. Nelson & Co Ltd,
5 Endeavour Way,
Wimbledon, London SW19 9UII
(081) 946 8527

Nelsons Homoeopathic Pharmacy,
73 Duke Street,
Grosvenor Square, London W1M 6BY
(071) 629 3118

Weleda UK Ltd,
Homoeopathic Pharmacy,
Heanor Road,
Ilkeston, Derbyshire DE7 8DR
(0602) 309319

*For a list of veterinary surgeons
practising homoeopathy contact:*
The Secretary,
The British Association of
Homoeopathic Veterinary Surgeons
(BAHVS),
Chinham House, Stanford-in-the-
Vale,
Faringdon, Oxfordshire SN7 8NQ
Please enclose an SAE.

HERBAL REMEDIES
Specifically Veterinary:
Denes Natural Pet Care,
PO Box 691,
2 Osmond Road, Hove,
East Sussex BN3 3SD
*Denes also supply additive-free
petfoods and offer an advisory service.*
(0273) 25364

Dorwest Herbs (Veterinary),
Shipton Gorge, Bridport,
Dorset DT6 4LP
(0308) 897272

OTHER HERB SUPPLIERS:
Gerard House Limited,
3 Wickham Road, Bournemouth,
Dorset BH7 6JX
(0202) 434116

Neal's Yard Remedies,
15 Neal's Yard, Covent Garden,
London WC2 9DP
(071) 379 7222

Potters Herbal Suppliers Ltd,
Leyland Mill Lane, Wigan,
Lancashire WN1 2SB
(0942) 34761

*For more information about herbs in
general contact:*
The Herb Society,
PO Box 599,
London SW11 4RW

BACH FLOWER REMEDIES
*These are usually available from good
health-food shops. In case of difficulty
contact:*
Bach Flower Remedies Ltd,
Mount Vernon,
Sotwell, Wallingford,
Oxfordshire OX10 0PZ
(0491) 39489

ESSENTIAL OILS
Most health-food shops will sell
essential oils, but be sure to obtain
those which are pure and natural.
Synthetic oils will not produce good
results and may cause harm.
Gerard House (see above) can supply
pure, natural essential oils.

**SUPPLEMENTS FOR VEGETARIAN
CATS**
Katz Go Vegan,
Box 161,
7 Battle Road, St Leonards-on-Sea,
East Sussex TN37 7AA

SUPPLIERS (USA)

HOMOEOPATHIC REMEDIES
Boericke and Tafel Inc,
1011 Arch Street,
Philadelphia, PA 19107

*To find out more about veterinary
surgeons practising homoeopathy in
the USA contact:*
Dr Stephen Tobin (IAVH),
26 Pleasant St,
Meriden, CT 86450

**HERBAL REMEDIES AND
ESSENTIAL OILS**
Should be available from local health-
food stores.

BACH FLOWER REMEDIES
Ellon (Bach USA) Inc,
PO Box 320,
Woodmere,
NY 11598
(516) 593 2206

**SUPPLEMENTS FOR VEGETARIAN
CATS**
Harbingers of a New Age,
9010 Lower Pack River Road,
Sandpoint,
ID 83864
(208) 263 7810

Index

Page numbers in *italics* refer to picture captions.

ACKNOWLEDGEMENTS

Editor	Michele Doyle
Copy-editor	Rosanne Hooper
Editorial Assistant	Zoë Hughes
Proof-reader	Sam Merrell
Art Director	Elaine Partington
Designer	Nigel Partridge
Illustrator	Danuta Mayer
Production	Hazel Kirkman and Charles James
Picture Researcher	Liz Eddison
Indexer	Dorothy Frame

Eddison Sadd Editions would like to thank the following, who gave permission to reproduce the illustrations:
t = top; b = bottom; l = left; r = right; c = centre

Animals Unlimited 18, 93; Miguel Arana/Trip 36; Joan Batten/Trip 20; Anne Marie Bazalik/Trip 8, 17, 44, 54, 56; Jane Burton 3, 21t,b, 24, 26, 38t,b, 40, 41, 43, 46t,b, 47r, 49, 52, 53, 58b, 59, 60, 61t,b, 67, 68, 76l,r, 77t,b, 79, 84b, 85l,r, 86, 87t,b; Jane Burton/Bruce Coleman 64-65, 66; Ed Buziak/Trip 2, 6, 51; Robert Estall 61c, 82, 88; Mike Evans/Trip 50; Marc Henrie 11, 14, 62; Julie Meech/Trip 91; Peter Saunders/Trip 96; Solitaire 29, 33t, 39, 42t,b, 47l, 58t, 74, 81, 84t, 92; Kim Taylor 33b; Sally Anne Thompson/Animal Photography 23; R. Willbie/Animal Photography 1.